Gambling 102

Gambling 102

The Best Strategies for All Casino Games

Second Edition

Michael "The Wizard of Odds" Shackleford

Huntington Press
Las Vegas, Nevada

Gambling 102
The Best Strategies for All Casino Games
Second Edition

Published by
Huntington Press
3665 Procyon St.
Las Vegas, NV 89103
Phone (702) 252-0655
e-mail: books@huntingtonpress.com

Copyright ©2019, Michael Shackleford
First Edition: ©2005

ISBN: 1-978-944877-18-7

Cover Photo supplied by Stockbyte™ Royalty Free Photos

Design & Production: Laurie Cabot

Contents

Acknowledgments

L et me start with those directly responsible for helping with this book. In 2000, I self-published 100 copies of a very rough draft of this book, titled *May the Odds Be with You*, and sent them to every known publisher of gambling books. Only one person showed any interest in a then-unknown name. Most didn't even give the courtesy of a rejection letter. That person was Anthony Curtis of Huntington Press. Without his faith in me, this book would not be in your hands right now.

Next, while I make no apologies for my math skills, let's just say my SAT score in English was 300 points less than that of my math score. Deke Castleman deserves 100% of the credit for taking this book from the grammatical mess it started as into the polished gem you see now. A fine team I think we made.

In addition, there would be no *Gambling 102* if it hadn't been for the encouragement and support of many. The math department at the Los Alamitos High School helped nurture my interest and skills in math and computer programming. Years later, in 2000, I nagged my wife to give her blessing to leaving a comfortable GS-14 government job to focus my energy on a website about gambling. Usually, people only leave such comfort and security for two reasons — retirement or death.

x • **Gambling 102**

Finally, I also thank my thousands of followers, like JB, mustangsally, Michael Bluejay, Mike Hopson, ThatDonGuy, Charlie Patrick, Daniel Dale, rsactuary, zcore13, Tringlomane, 7craps, and miplet, who keep me on my toes by providing feedback and sometimes correcting mistakes. Everything you read in this book has already been peer reviewed by the world on my website and those peers deserve much credit for this book and my success.

Foreword

The dictionary tells us that a *wizard* is: "A very clever or skillful person who manages to do something that is extremely difficult; a person worthy of the highest praise."

This is the perfect description of Michael Shackleford, the self-proclaimed "Wizard of Odds." It has been my privilege to know Mike for many years, and his reputation in the gaming industry is impeccable. Quite simply, when you want the most reliable, accurate, and complete analyses of the mathematics of virtually any gambling game in today's ever-expanding market, Shackleford has to be your first, go-to, and most authoritative source.

With the new and original material in this second edition of *Gambling 102*, Mike brings up to date his first effort to explain, in his typical straightforward no-nonsense style, the intricacies of today's most popular casino games. As he writes in the Introduction, this is not a beginner's reference to the rules or a general orientation to how the games are played. Instead, the current volume provides detailed expositions on the odds and house edges of nearly 20 casino offerings, often with a variety of versions and variations within the main game. The result is the most trustworthy compendium of gaming odds to be found anywhere in print.

To this already impressive scholarship, Mike brings a new metric, the "Element of Risk," designed to measure and compare not simply the traditional house edges, but those expectations adjusted for the amount of "action" per resolved bet. The result is a more accurate, and completely original, approach to determining which games are truly the "best" to play.

Too many gambling books rehash the standard Gambling 101 materials that are all too familiar. With *Gambling 102*, we enjoy a completely novel methodology applied to these same games, which permits us to consider them in a different, more meaningful, manner. Mike "The Wizard of Odds" Shackleford is simply the very best at what he does. Enjoy this delightful new work from a true master of the craft.

—Don Schlesinger, author of
Blackjack Attack: Playing the Pros' Way

Introduction

On my website WizardOfOdds.com, I've analyzed approximately 400 casino games, 160 video poker games, 50 blackjack side bets, 40 baccarat side bets, hundreds of side bets in other games, and so many rule variations of popular games that I've lost count. This is all the product of 22 years of analyzing and writing about casino games.

However, not every recreational gambler needs or wants to know about a game that crashed and burned in a field trial more than a decade ago.

For advanced recreational gamblers who already know the rules to the games and want to know the house edge they face and the proper strategy to play, this book is for you.

There's a reason the title is *Gambling 102*, as opposed to *Gambling 101*. Plenty of books on gambling for beginners already exist. This book, by contrast, endeavors to build on a foundation of the basics by presenting a more rigorous look at the odds. As such, the rules of the games are rarely explained within. If you don't know the difference between blackjack and baccarat, the information here will probably be too advanced. If you're already a skilled advantage player in a range of games, you probably already know most of what's presented. I wrote this book for those of you who are somewhere in the middle.

The first edition of *Gambling 102* was published in 2005. The casino floor is always changing. Old games die out and new ones come along. Gone from this edition are chapters on Caribbean Stud Poker, Casino War, big six, racetrack betting, and Texas Hold 'em.

Added to this edition are popular new table games that have come along since 2005: Crazy 4 Poker, High Card Flush, Mississippi Stud, Texas Hold 'em Bonus, and Ultimate Texas Hold 'em.

One of the new chapters of which I'm most proud is on pai gow tiles, which has been around for hundreds of years in China. I just didn't know enough about it in 2005 to include a proper chapter. This second edition also includes a more general chapter on sports betting, including coverage of the NFL, MLB, and NBA, and a big section on video keno.

It is my goal that this book not only helps you win, or lose less, in the casinos, but also conveys an appreciation of the math behind the games.

If you play for the other side, I also hope you learn about the math behind the games you're dealing or supervising.

I view every casino game as a math question and *Gambling 102* is the answer book. As long as my readers are learning something, I'm happy, and it doesn't matter which side of the table you're on.

A Word About the Bell Curve

I've always struggled with what to do in situations where most people say, "Good luck." I don't believe in luck, at least as defined as some kind of mystical force that causes good and bad things to happen.

In the casino, as well as life in general, what happens to

us is a function of the decisions we make and random chance. If you smoke (preferably not next to me at the pai gow table), the earlier you can expect to die. Likewise, if you play slot machines and make sucker bets at the tables, the more money you can expect to lose. Both are based on averages, not guarantees on an individual basis.

However, neither are certainties. In the casino, how much money you win or lose depends on three factors:

- game and bet selection
- for games of skill, the decisions made playing
- randomness

In the short run, randomness plays the biggest part. In the long run, the ups and down of randomness mostly cancel out, rendering game selection and skill the greatest factors of how well you do.

After the dust clears from all the math, the net win or loss from a number of bets can be very closely approximated by throwing a dart at a "bell curve." The shape of a bell curve has just two factors, the mean and variance. In other words, where is it centered and how wide is it? To be specific, the following equation shows the probability of any outcome, x, given mean μ and standard deviation σ, on the bell curve:

$$\frac{1}{\sigma\sqrt{2\pi}}\, e^{-\frac{1}{2}(\frac{(x-\mu)^2}{\sigma})^2}$$

What most people call luck comes down to the formula above. Note how it beautifully captures both transcendental numbers e and e, much like Euler's equation, $e^{\pi i} + 1 = 0$.

While no book can help where that dart lands on the bell curve, this one strives to make the mean result μ, as high as possible. It also advises on the σ, depending on whether the player is seeking a low- or high-volatility game. In other words, you won't win every bet based on the advice in this book, but if you follow it, you can expect to lose much less, or even win, over the long run, depending on which game you play and how well you play it.

Getting back to my dilemma of what to say in lieu of "good luck," I prefer Effie Trinket's line from *The Hunger Games*, "May the odds be ever in your favor."

House Edge and the Element of Risk

A common statistic bandied around for casino games is the house edge. It's an important number for players, in that it tells them how much it costs to play a game. For example, over time, you'll lose $5.26 for every $100 wagered at double-zero roulette, compared to just $1.41 per $100 on the pass line in craps. The house edge is equally important for the casinos, both to calculate the earn and to quantify the value of a gambler's play. The formula for the latter is simple: expected casino win = house edge x average initial bet x bets per hour x hours played.

All good. But where the waters get muddy is when players and other gambling writers use the house edge as an absolute measure of a game's value. Using the house edge is fine for simple games like roulette that have no pushes and no supplemental wagers. However, for games with supplemental wagering, the house edge is a bad measure of game value. Instead, I like to use a statistic that I call the "element of risk" (EOR).

Before I get into the EOR, let's clarify that my definition of the house edge is the expected player loss relative to the original bet made. This includes pushes. Some older gambling books define the house edge as the expected player loss per resolved bet. This makes a difference in bets like the don't pass in craps and the banker bet in baccarat. Most modern gambling writers, including me, count a tie as a resolved bet. The only possible exceptions worth mentioning are multi-roll bets that can be taken down in craps, e.g., place and hardway bets. Some argue that if a single roll does not resolve such bets, then the outcome is a push. I don't. I assume the bet stays up until resolved for all crap bets, except the don't pass and don't come, where I consider a push to be a resolved bet. The two different ways of defining a resolved bet account for some slight differences you may see in house-edge numbers from different credible sources.

That said, where applicable, in my opinion, the element of risk is the superior gauge. I define element of risk as the expected player loss divided by the average amount bet by the end of the hand. A good example of a game where this is significant is Three Card Poker. The traditionally calculated house edge in that game is 3.37%, measured as the ratio of the expected player loss to the Ante bet. However, the player raises 67.42% of the time, for an average final bet of 1.6742 units. The element of risk is thus 3.37%/1.6742 = 2.01%. In other words, the player can expect to lose 2.01% of all money bet, if he avoids all the side bets and plays under the usual 1-4-5 Ante Bonus paytable.

In this book, for games where there's supplemental wagering, I provide both the house edge and element of risk. In games where there isn't, I list the house edge only, because the element of risk is the same.

I hope this wasn't too confusing, but there's already much confusion about the definition of the house edge and I'm trying to clear up the problem by introducing this new statistic.

1

The Ten Commandments of Gambling

Thou shalt not cheat.
No explanation necessary.

Thou shalt honor thy gambling debts.
A true gentleman honors his debts, especially gambling debts. When making a bet with another person, you're putting your honor on the line. If you lose, you pay. No excuses!

Thou shalt expect to lose.
The Las Vegas Strip wasn't built because people usually win in the casinos. Even with good rules and strategies, the odds are still usually in the casino's favor. So don't get mad if you lose. Think of it as the price you pay for entertainment.

Thou shalt trust the odds, not hunches.
If you want to maximize your odds, believe in mathematically proven strategies, such as those in this book, rather than hunches.

Thou shalt not over-bet thy bankroll.
Before you start playing, set a gambling budget for yourself and stick to it. Don't gamble with money you need for necessities.

Thou shalt not believe in betting systems.

For every one legitimate gambling writer, 100 charlatans are trying to sell worthless betting systems that promise an easy way to beat the casinos. I know it's a cliché, but if it sounds too good to be true, it probably is.

Thou shalt not hedge thy bets.

Hedge bets usually carry a high house edge. For example, never take insurance in blackjack and never bet the any craps or any seven in craps. Exceptions can be made for insuring life-changing amounts of money.

Thou shalt covet good rules.

Rules vary from casino to casino. To improve your odds, know the good rules from the bad, then seek out the best rules possible.

Thou shalt not make side bets.

Side bets are sucker bets. Period.

Thou shalt have good gambling etiquette.

Gambling is a lot more fun when people are polite and respect one another. It's also proper etiquette to tip for good service.

2

Casino Etiquette

The same standards for etiquette outside the casino also apply inside. Emotions can run the whole gamut in a casino, but regardless of whether you're on an extreme high, extreme low, or anywhere in between, you should still exercise common sense, restraint, and respect for others.

That said, here are some etiquette tips that are specific to casinos.

- Don't correct or critique another player's play unless he's receptive to suggestions. In particular, don't harass the last player to act in blackjack. It's a myth that a bad player, particularly at third base, causes the entire table to lose.
- Be courteous to the dealers. It isn't their fault when you lose. If you can't take losing, don't play at all.
- Tip the dealers. Dealers usually make minimum wage and rely on tips to make a decent living. Losing is not an excuse not to tip. Dealers should be tipped according to the level of service they provide.
- Tip the cocktail waitresses. One dollar per drink is the minimum. Ordering water does not excuse you from tipping.
- Tip slot personnel when you receive a hand-pay jack-

pot. The amount is touchy. My personal opinion, which I admit is on the low side, is 0.5% to 2%, depending on the size of the jackpot; the larger the jackpot, the lower the percentage. Tipping more seems to be expected at small (non-casino) bars.

- Do your best to understand rules and strategy before you play. Try not to slow down a game because you don't know how to play, unless you're the only one playing.

- If you make a bet using different chip denominations on a table game, put the higher denominations on the bottom of the stack.

- Once you make a bet at a table game, never touch it unless you win or push.

- Don't drink past the point where it annoys other people.

- I recognize that smokers, unfortunately, have a right to smoke in casinos. However, if you must, try to be respectful of non-smokers, especially if they were at the table first. Ideally, if there are non-smokers at the table, it would be nice if you stepped away from the table to smoke.

- If you enter a blackjack game in the middle of the shoe, ask the other players if you may join mid-shoe and do so only if there are no objections. In a single- or double-deck game, I would just wait for the shuffle, which is often required anyway.

- Don't hog multiple betting spots if other players are waiting to play.

3

Baccarat

For players seeking an extremely easy game with a low house advantage, it's hard to beat baccarat. The rules and ambience may seem intimidating, but all you need to know to get started is bet selection. There are no other decisions to be made and the dealer controls the cards according to rules that you don't need to understand.

The two main bets in baccarat are the Player and Banker. I put them in capital letters to avoid confusion. Players can also bet on a tie and usually a Player pair and Banker pair.

The following table tells you what you really need to know about baccarat. For those five major bets, the table shows the probability of a win, tie, loss, and most importantly, the house edge.

TABLE 1—Baccarat Probability and House Edge

Bet	Pays	Prob. Win	Prob. Tie	Prob. Loss	House Edge
Player	1	44.63%	9.51%	45.87%	1.24%
Banker	0.95	45.87%	9.51%	44.63%	1.06%
Tie	8	9.51%	0.00%	90.50%	14.44%
Player pair	11	7.47%	0.00%	92.53%	10.36%
Banker pair	11	7.47%	0.00%	92.53%	10.36%

(Some of the horizontal percentages don't add up to 100% due to rounding.)

Strategy

The strategy is simple—bet the Banker every time. Every other player will be looking for patterns in the history of wins and losses. Let them waste their time; it doesn't help. However, if bouncing back and forth makes the game more fun for you, go ahead, with the understanding that the odds on the Player bet are 0.18% worse. Just avoid the tie, pairs, and any other side bet the casino may add.

Alternate Pays

Some casinos have offered a 4% commission on the Banker bet, as opposed to the traditional 5%. This lowers the house edge on the Banker to 0.60%.

The Tie bet has also been seen, though rarely, to pay 9-1. This lowers the house edge on the Tie to 4.93%.

Commission-Free Baccarat

In an effort to speed up the game, more and more casinos are paying even money on the Banker bet. However, there's always some take-back when a rule change cuts into the edge of the player's way. I've heard of several versions, but the following two are the most popular.

EZ Baccarat—A Banker 3-card total of 7 is a push. The house edge is a bit less than conventional baccarat at 1.02% on the Banker bet. A side bet (called the Dragon) on a 3-card winning Banker total of 7 pays 40-1 and has a probability of winning of 2.25% and a house edge of 7.61%. A side bet called the Panda 8 pays 25-1 on a winning 3-card Banker 8. The

Panda 8 has a probability of winning of 3.45% and a house edge of 10.19%.

Nepal Baccarat — A Banker winning total of 6 pays 1-2. The house edge on the Banker bet is higher than standard baccarat at 1.46%. Different side bets win on a Banker winning total of 6. One of them, the Lucky 6, pays 20 if the winning Banker 6 is composed of three cards and pays 12 if composed of two cards. The bottom line on the Lucky 6 is a house edge of 16.68%.

Egalités

Egalités are a set of 10 side bets, one on each tie of 10 possible ties. I've seen these bets online and they're the norm at casinos in the UK. The pays vary. The following table shows a common paytable.

TABLE 2—Egalités Probability and House Edge

Egalite	Pays	Prob. Win	House Edge
9-9	80	1.10%	10.63%
8-8	80	1.10%	11.07%
7-7	45	2.04%	6.39%
6-6	45	1.92%	11.50%
5-5	110	0.79%	11.87%
4-4	120	0.73%	12.14%
3-3	200	0.45%	10.52%
2-2	225	0.40%	9.54%
1-1	215	0.41%	11.42%
0-0	150	0.58%	12.45%

Dragon Bonus

The Dragon Bonus is a pair of side bets, one on the Player and one on the Banker, that win if the chosen side wins with a natural (a two-card 9) or by at least four points. The only paytable I've ever seen is 30, 10, 6, 4, 2, 1, 1. The bottom line is a house edge of 2.65% on the Player Dragon and 9.37% on the Banker Dragon.

Other Side Bets

My website lists 33 different side bets for baccarat and commission-free baccarat. I've addressed four of the most popular here. Space doesn't allow me to cover all the others. As with any game, my advice when you see a side bet with which you're not familiar is, with all due credit to Nancy Reagan, just say "no."

4

Blackjack

Blackjack, if played properly, is one of the best bets in the casino. Just a basic strategy player can often get back more in comps than he gives to the casino via its thin house advantage.

This chapter merely scrapes the surface of a game about which I could easily write a college textbook. The two most important considerations for beginners are game selection and how to properly play any given hand. Following a brief discussion of game selection, I present three strategies that will take you from beginner to a level that will put you in the top 1% of blackjack players.

Game Selection

Half the battle in blackjack is choosing a table with good rules. Most casinos have an assortment of tables and rule sets. In general, the better rules are found at the tables with higher minimums, but there are many exceptions. It always pays to survey any casino for the best rules at minimums you're comfortable with before sitting down to play.

The single most important rule is that a winning blackjack (or "natural") pays at 3-2 odds. If it doesn't, it should be a deal breaker.

That said, let's consider a starting point of an 8-deck game where the dealer hits a soft 17, a cut card is used, and there are no other player-friendly options. Even with these lousy rules, the house edge is still a low 0.81%, better than most casino games. From there, take a look at the following list of common player-friendly rule variations and their effect on the house edge. A negative effect means a drop in the house edge, so 1 deck is the best variation, removing nearly half a percentage point, while a continuous shuffler is the least, though it still makes a (tiny) difference.

TABLE 3—Effect of Good Rules on Blackjack Edge	
Rule	**Effect**
1 deck	–0.50%
Dealer stands on soft 17	–0.22%
2 decks	–0.20%
Surrender allowed	–0.09%
Re-splitting aces allowed	–0.08%
6 decks	–0.02%
Continuous shuffler	–0.01%

The takeaway from the table above is that the fewer the decks, the better the odds for the player. Given a choice, it's better for the player if the dealer stands on soft 17. Adding options, like surrender and re-splitting aces, is always good. Finally, yes, for the recreational player, the odds are a little better with a continuous shuffler, compared to a cut-card game.

Let me also address some player-unfriendly rules you may encounter and how much they increase the house edge.

TABLE 4—Effect of Bad Rules on Blackjack Edge	
Rule	**Effect**
Blackjack pays 1-1	+2.27%
Blackjack pays 6-5	+1.36%
Player may double on hard 10 or 11 only	+0.18%
No dealer hole card	+0.11%
Player may double on hard 9-11 only	+0.10%

Again, if there's just one thing to remember from the table above, it's that the effect of the house's paying less than 3-2 on a blackjack far overshadows any other variable. Don't be fooled by blackjack machines that say a blackjack pays 2-*for*-1; that's the same thing as 1-*to*-1 (even money).

Basic Strategy

The basic strategy chart on the following page is optimized for a game of 4 or more decks in which the dealer hits a soft 17. There are different basic strategies for other sets of rules, but the changes are relatively minor. If you want to play perfectly, the more precise basic strategies for most games can be found at WizardOfOdds.com. (The best consolidated presentation of basic strategies for almost every conceivable combination of game conditions is presented in the book *The Theory of Blackjack* by Peter Griffin.)

Given that at most tables, the dealer hits soft 17 and deals from a 6- to 8-deck shoe, I believe that if you memorize just one basic strategy, it should be this one.

Player		Dealer									
		2	**3**	**4**	**5**	**6**	**7**	**8**	**9**	**10**	**A**
Hard	8	H	H	H	H	H	H	H	H	H	H
	9	H	D	D	D	D	H	H	H	H	H
	10	D	D	D	D	D	D	D	D	H	H
	11	D	D	D	D	D	D	D	D	D	D
	12	H	H	S	S	S	H	H	H	H	H
	13	S	S	S	S	S	H	H	H	H	H
	14	S	S	S	S	S	H	H	H	H	H
	15	S	S	S	S	S	H	H	H	Rh	Rh
	16	S	S	S	S	S	H	H	Rh	Rh	Rh
	17	S	S	S	S	S	S	S	S	S	Rs
Soft	A,2	H	H	H	D	D	H	H	H	H	H
	A,3	H	H	H	D	D	H	H	H	H	H
	A,4	H	H	D	D	D	H	H	H	H	H
	A,5	H	H	D	D	D	H	H	H	H	H
	A,6	H	D	D	D	D	H	H	H	H	H
	A,7	Ds	Ds	Ds	Ds	Ds	S	S	H	H	H
	A,8	S	S	S	S	Ds	S	S	S	S	S
Pairs	2,2	Ph	Ph	P	P	P	P	H	H	H	H
	3,3	Ph	Ph	P	P	P	P	H	H	H	H
	4,4	H	H	H	Ph	Ph	H	H	H	H	H
	5,5	D	D	D	D	D	D	D	D	H	H
	6,6	Ph	P	P	P	P	H	H	H	H	H
	7,7	P	P	P	P	P	P	H	H	H	H
	8,8	P	P	P	P	P	P	P	P	P	Rp
	9,9	P	P	P	P	P	S	P	P	S	S
	T,T	S	S	S	S	S	S	S	S	S	S
	A,A	P	P	P	P	P	P	P	P	P	P

Key:
H = Hit
S = Stand
D = Double if allowed, otherwise hit
Ds = Double if allowed, otherwise stand
P = Split
Ph = Split if double after split allowed, otherwise stand
Rh = Surrender if allowed, otherwise hit
Rs = Surrender if allowed, otherwise stand
Rp = Surrender if allowed, otherwise split

Again, never take insurance. This includes declining "even money."

Wizard's Simple Strategy

In trying to teach friends and relatives to play blackjack for more than 30 years, I've come to the conclusion that memorizing the complete basic strategy is too much to ask of most people. For those of you who want to improve your blackjack game with minimum effort, I present the following strategy. It groups player hands into 11 different categories and only two types of dealer up-card. It yields the correct play most of the time and for those where it doesn't, the error is marginal. All things considered, the increase in house edge due to errors using this strategy is only 0.14%. That's about one hand in every 12 hours of play.

Player's Hand		Dealer's Up-Card	
		2 to 6	7 to A
Hard	4 to 8	H	H
	9	D	H
	10 or 11	Double with more than dealer	
	12 to 16	S	H
	17 to 21	S	S
Soft	13 to 15	H	H
	16 to 18	D	H
	19 to 21	S	S
Splits	22,33,66,77,99	P	N
	88,AA	P	P
	44,55,TT	N	N
Key: H = Hit S = Stand D = Double P = Split N = Don't split			

Additionally:

- Surrender 16 vs. 10, if allowed.
- If the strategy indicates a soft double, but that isn't allowed, hit with soft 17 or less and stand with soft 18.
- If you have a pair and the strategy says don't split, refer to the hard-total strategy.
- With a total of 10 or 11, double if your total is higher than the dealer's up-card, counting a dealer ace as 11.
- Never take insurance.

Wizard Ace-Five Count

If you're looking for an easy way to get the odds in your favor, then consider my Ace-Five Count to be an easy jump up from the basic strategy and a first step to card counting.

Using it is quite simple, as follows:

- Find a cut-card game dealing 6 or more decks with as good rules and penetration (how deep the cut card is placed) as possible.
- Set a bet spread you're comfortable with, where the maximum bet is eight times the minimum. For example, $10-$80.
- After a shuffle, start the count at 0 and your bet at the minimum of your range.
- For every 5 that's played, add 1 to the count.
- For every ace that's played, subtract 1 from the count.
- If the count is +2 or greater, double your last bet (until you reach the maximum of your bet range).
- If the count is +1 or less, revert to your minimum bet.
- Play every hand by the basic strategy.

There you have what is, as far as I know, the easiest-to-use card-counting strategy out there. No true-count conversions, no index numbers to memorize, and only two ranks to count. This isn't, to be sure, a strategy meant for professional counters, but if you want to easily swing the odds in your favor, albeit very marginally, then welcome to the world of advantage play. You can always graduate to more powerful card-counting strategies later.

Using the Ace-Five Count cuts the house edge by about 0.56% using a spread of 1-8, depending on other factors, mainly the penetration. If you increase the maximum wager to 16 or 32, you'll increase the power even more. You can find more information about this count's effectiveness on my website.

Side Bets

At the time of this writing, my website has analyzed 42 different blackjack side bets. Let me summarize all 42 of them in two words: sucker bets. That summary goes for side bets in all games, not just blackjack.

5

Craps

Craps seems like an intimidating game from the outside. It moves very fast and has a multitude of bet options, as well as a language all its own. However, the essentials of craps are quite simple. Like all other casino table games, you need to know only one bet to play, though understanding the others will increase your enjoyment. Most players make the same basic bets; this parallel action results in the excitement often seen at the table when the dice are rolling well. It can provide for a nice contact high at a crowded table that's winning. Best of all, the house advantage is among the lowest in the casino, if you stick to the main bets.

Strategy

To achieve the best possible odds, follow these two steps:

- Bet the pass, don't pass, come, or don't come.
- If the bet is not resolved on the first throw, take the maximum allowed odds behind it.

I would be remiss in my duties if I didn't mention that the odds are slightly better on the don't pass and don't come than on the pass and come. However, most players bet with the dice on the pass and come, rather than against them. If it makes the game more fun for you to bet with the table, go right ahead. The difference in the house edge is marginal.

Multi-Roll Bets

Many bets in craps take multiple rolls to resolve. The following table shows the house edge (the expected loss per bet resolved) on all of them.

TABLE 5—House Edge on Multi-Roll Bets		
Bet	**Pays**	**House Edge**
Pass, come	1-1	1.41%
Don't Pass, don't come	1-1	1.40%
Odds on a 4 or 10	2-1	0.00%
Odds on a 5 or 9	3-2	0.00%
Odds on a 6 or 8	6-5	0.00%
Odds against a 4 or 10	1-2	0.00%
Odds against a 5 or 9	2-3	0.00%
Odds against a 6 or 8	5-6	0.00%
Buying 4 or 10*	13-7	4.76%
Placing 5 or 9	9-5	2.78%
Placing 6 or 8	7-6	1.52%
Big 6 or 8	1-1	9.09%
Laying the 4 and 10*	19-41	2.44%
Laying the 5 and 9	19-31	3.23%
Laying the 6 and 8	19-25	4.00%
Hard 4 and 10	7-1	11.11%
Hard 6 and 8	9-1	9.09%

*The odds expressed assume the commission must be pre-paid. However, some liberal casinos charge the commission on buy and lay bets on the 4 and 10 only if they win. This lowers the house edge on the 4 and 10, both buying and laying, to 1.67%.

Between place and buy bets, I've indicated only the better of the two. At the table, buy and lay bets are expressed as fair odds, minus a 5% commission; for expedience, I've converted that to a single ratio.

Blended House Edge

The following table shows the overall (blended) house edge when playing the pass/don't pass and buying/laying the maximum allowed odds. The maximum ratio of the odds bet to line bet varies from casino to casino.

TABLE 6—House Edge for Various Odds Multiples

Odds Allowed	Pass and Buying Odds	Don't Pass and Laying Odds
1X	0.85%	0.68%
2X	0.60%	0.45%
3-4-5X*	0.37%	0.27%
3X	0.47%	0.34%
5X	0.33%	0.23%
10X	0.18%	0.12%
20X	0.10%	0.06%
100X	0.02%	0.01%

*3-4-5X odds assumes the player can bet 3X on a point of 4 or 10, 4X on a point of 5 or 9, and 5X on a point of 6 or 8. This works out to a win of 6X the pass or come bet for all points, assuming full odds are taken.

Single-Roll Bets

Craps offers plenty of ways to throw your money away on bets that resolve after one roll. You'll see all these sucker bets, except the over/under 7, conveniently grouped in the middle of the crap table. The following table shows the house edge of all the single-roll bets. As you can see, the payout for some bets can vary. The most liberal odds can be found in the UK and Australia.

TABLE 7—House Edge on Single-Roll Bets

Bet	Pays	House Edge
2, 12, all hard hops	29	16.67%
2, 12, all hard hops	30	13.89%
2, 12, all hard hops	31	11.11%
2, 12, all hard hops	32	8.33%
2, 12, all hard hops	33	5.56%
3, 11, all easy hops	14	16.67%
3, 11, all easy hops	15	11.11%
3, 11, all easy hops	16	5.56%
7	4	16.67%
7	4.5	8.33%
Any craps	7	11.11%
Any craps	7.5	5.56%
Over/Under 7	1	16.67%

Field

While the field is a single-roll bet, the payouts vary depending on the total thrown. The stingy rules pay 2-1 on both a 2 and 12, for a house edge of 5.56%. The liberal rules pay 3-1 on either the 2 or 12, but not both, for a house edge of 2.78%.

Side Bets

One can easily argue that basic craps already has lots of side bets, but some new ones have been added to the game that allow for decisions based on a shooter's entire set of rolls. The two most common are the Fire Bet and Bonus Craps.

The Fire Bet pays according to how many different unique points the shooter makes before a seven-out. Following are three paytables I'm aware of and the house edge of each.

25, 250, 1000 — 20.76%
10, 200, 2000 — 24.86%
7, 30, 150, 300 — 20.73%

Bonus Craps is actually three side bets that win if the shooter rolls every total in the given range before rolling a 7. The Small bet usually pays 30-1 for a house edge of 18.30%. The Tall bet usually pays 150-1 for a house edge of 20.61%.

Crapless Craps

This is a variation that was developed at the old Vegas World in Las Vegas. Its successor, the Stratosphere, now known as the Strat, continues to deal the game, which also shows up from time to time in other places. While a novelty in Las Vegas, Crapless Craps is the norm in Detroit casinos.

On the pass line bet, a 7 on the comeout roll still wins; however, any other number becomes the point, including the 2, 3, 11, and 12. At first, this may seem like a positive change, because the player now has hope on the 2, 3, and 12 in exchange for giving up the sure winner on the 11. The problem is the 2, 3, and 12 will likely lose anyway, while the 11 changes from a winner to a likely loser. Overall, the house edge is 5.38% on the pass line, almost four times as high as conventional craps.

6

Crazy 4 Poker

Crazy 4 Poker is one of the most successful of the wave of poker-based games to come out after Three Card Poker. The game has elements of both Three Card Poker and Ultimate Texas Hold 'em. It's like Three Card Poker in putting the player to a raise or fold decision. It's like Ultimate Texas Hold 'em in allowing big raises for premium hands at the cost of mandating the Super Bonus bet, which is no different than a very bad side bet.

It's hard to beat Crazy 4 Poker's simple strategy and cost of playing (a little more than 1% of money bet). In fact, I can't think of a game with better odds that has a simpler strategy. Craps would be the only contender.

Analysis

The base game involves Ante, Raise, and Super Bonus bets, which are all interconnected. To evaluate the value of the overall game, you have to consider the value of each bet, which is shown in the following table.

TABLE 8—Crazy 4 Poker House Edge		
Bet	**Average Wager**	**Return**
Ante	1.000000	–0.171298
Raise	1.136171	0.484904
Super Bonus	1.000000	–0.347795
Total	**3.136171**	**–0.034189**

The lower-right cell shows that if you start with initial bets of one unit each on the Ante and Raise, you can expect to lose 0.034189 by the end of the hand. If we define the house edge as the ratio of the expected loss to the Ante bet, that's 3.42%. However, using that statistic as a measure of the game's player value is deceptive. By the end of the hand, you'll have bet a total of 3.136171 units, adding up the Ante, Raise, and Super Bonus. That makes the element of risk equal to 3.42%/3.14 = 1.09%, which is quite low compared to most games.

Strategy
The strategy for Crazy 4 Poker is quite simple:

- If you can make the large raise, do so.
- If you have K-Q-8-4 or better, make the small raise.
- Otherwise, fold.

Queens Up
Of course, Crazy 4 Poker comes with at least one side bet. The one you always see is the Queens Up. It's called that because the player needs at least a pair of queens to win. The

following table shows four known paytables. Naturally, most casinos opt for Table 4, which has the highest house edge.

TABLE 9—Queens Up Side Bet House Edge				
Hand	Table 1	Table 2	Table 3	Table 4
4-of-a-Kind	50	50	50	50
Straight Flush	30	40	30	40
3-of-a-Kind	9	8	8	7
Flush	4	4	4	4
Straight	3	3	3	3
Two Pair	2	2	2	2
Pair of Queens or Better	1	1	1	1
House Edge	3.06%	4.52%	5.32%	6.78%

6-Card Bonus/Millionaire Maker

This side bet is common to many ShuffleMaster games. There are several versions and paytables, but with a house edge ranging from 9% to 18%, let's just file this one under "all side bets are sucker bets."

7

Face Up Pai Gow Poker

The name of the game, Face Up Pai Gow Poker, describes it quite well. Not only are all the dealer cards exposed, but there's no 5% commission.

You might ask, "What's the catch?" First, if the dealer's best 5-card hand is an ace-high, the outcome is an automatic push. This is a lousy hand in Pai Gow Poker, which would otherwise likely result in a player win. Second, there's no player banking.

If you're looking for a slow table game with minimal volatility and no skill required, Face Up Pai Gow Poker might the perfect game for you.

House Edge

Following are the possible outcomes in Face Up Pai Gow Poker.

TABLE 10—Face Up Pai Gow Possible Outcomes	
Win	24.79%
Tie	48.62%
Loss	26.59%

The house edge in this even-money game is simply the probability of losing minus the probability of winning, which comes to 1.81%, better than the higher 2.72% house edge in Pai Gow Poker, assuming the player follows the house way and never banks.

In the hundreds of casino games I've analyzed, I believe this game to be the least volatile, with a probability of a push of almost 50%. The standard deviation is a very low 0.72. In Pai Gow Poker and pai gow tiles, it's a bit higher at 0.75. You get a good bang for your buck in this game.

Strategy

With the dealer cards dealt face up, there's no strategy — other than not screwing up. There are 21 ways you can choose two cards out of seven for the low hand. If it's possible to win, set any hand that does. If it's possible to push, set any hand that does. Otherwise, throw in the towel.

Side Bets

Naturally, a side bet covers a dealer ace-high. I'm familiar with two paytables, as follows:

TABLE 11—Ace-High Side Bet House Edge		
EVENT	**Paytable 1**	**Paytable 2**
Player and dealer ace high	40	40
Dealer ace high with joker	15	12
Dealer ace high without joker	5	5
House Edge	**9.29%**	**10.52%**

Conclusion

If you're looking for a slow table game, with minimal volatility and no skill required, Face Up Pai Gow Poker may the perfect game for you.

8

High Card Flush

High Card Flush is an easy-to-play poker variant where the object is to have a longer flush than the dealer's. It was introduced at Harrah's Laughlin in 2011 and slowly grew from there to many placements at the time of this writing.

The structure is similar to Three Card Poker or Caribbean Stud Poker, where the player is put to a raise-or-fold decision after seeing his cards. The size of the raise depends on the player's cards: The better the cards, the greater the raise allowed. With optimal strategy, the element of risk is a fairly low 1.54%.

This is a game of skill that has a very easy strategy at competitive odds. However, I should add that Crazy 4 Poker also offers a very simple strategy at an element of risk of 1.09%, compared to the 1.54% in High Card Flush.

Analysis

The following table shows the probability of and contribution to the return under optimal strategy.

TABLE 12—High Card Flush Probability and House Edge

Event	Win	Probability	Return
Player raises 3X, dealer qualifies, player wins	4	0.001618	0.006473
Player raises 2X, dealer qualifies, player wins	3	0.021472	0.064417
Player raises 1X, dealer qualifies, player wins	2	0.258181	0.516361
Player raises, dealer does not qualify	1	0.167103	0.167103
Push	0	0.000840	0.000000
Fold	−1	0.321365	−0.321365
Player raises 1X and loses	−2	0.228857	−0.457715
Player raises 2X and loses	−3	0.000560	−0.001679
Player raises 3X and loses	−4	0.000003	−0.000013
Total		**1.000000**	**−0.026418**

The lower-right cell of the table above shows a house edge, defined as the ratio of the expected loss to the initial bet, of 2.6418%. The player will raise 67.9% of the time and the average raise, when made, is 1.05 units. Overall, the average bet by the end of the hand is 1.71 units. This makes the element of risk 1.54%.

Strategy

With a 5-card or more flush, always make the maximum allowed raise.

Optimal strategy is clear most of the time, as follows: With a 3-card flush of J-9-6 or higher, raise; and with a 3-card flush of 9-7-4 or lower, fold.

What about those hands in the middle, between 9-7-5 and J-9-5? Here the strategy gets more complicated and depends on the other four cards.

However, the basic strategy, created by Charles Mousseau, that ignores the other four cards that aren't part of the longest 3-card suit, says to raise with 10-8-6 or higher. Following this strategy results in a house edge of 2.71%, which is 0.07% higher than optimal strategy.

If you desire to get closer to optimal, please see the Gordon Michaels' strategy I have at WizardOfOdds.com, which cuts the cost of errors to 0.04%. Sorry, but it's too long and esoteric to include here.

Flush Side Bet

The Flush side bet pays based on the longest flush in the player's hand and does not require the player to beat the dealer. There are two known paytables as follows:

TABLE 13—Flush Side Bet House Edge		
Longest Flush	Paytable 1	Paytable 2
7 cards	300	300
6 cards	100	75
5 cards	10	5
4 cards	1	2
House Edge	7.81%	7.53%

Straight Flush Side Bet

The Straight Flush side bet pays based on the longest straight flush in the player's hand and also does not require the player to beat the dealer. There is one paytable only that I am aware of, as follows.

TABLE 14—Straight Flush Side Bet House Edge	
Longest Straight Flush	**Pays**
7 cards	8000
6 cards	1000
5 cards	100
4 cards	60
3 cards	7
House Edge	**13.11%**

9

Mississippi Stud

Mississippi Stud can be described as a variant of Let It Ride. In both games, the player is paid based on the poker value of a 5-card hand. Unlike Let it Ride, Mississippi Stud has three decision points. The choice at each is to make a small raise, large raise, or fold. The strategy isn't difficult to memorize and if played properly, Mississippi Stud is a top-ten game in terms of the best odds in the casino.

Odds

The ratio of money the player can expect to lose to his Ante bet is 4.91%, following the typical 2-3-4-6-10-40-100-500 paytable. That may sound average, but after the Ante bet, the player will put up 3.59 as much in raises, on average. This makes the element of risk 4.91%/3.59 = 1.37%, which is better than most games.

Strategy

The following strategy is optimal for the paytable mentioned above. A "made hand" is one that will at least push. Some situations refer to the number of points in the player's hand. Following is how many points each card is worth.

J-A   = 2 points
6-10 = 1 point
2-5   = 0 points

2 Cards

Raise 3X with any pair.
Raise 1X with at least two points.
Raise 1X with 6,5 suited.
Fold all others.

3 Cards

Raise 3X with any made hand.
Raise 3X with a royal flush draw.
Raise 3X with a straight flush draw with no gaps, 5,6,7 or higher.
Raise 3X with a straight flush draw with one gap and at least one high card.
Raise 3X with a straight flush draw with two gaps and at least two high cards.
Raise 1X with any other three suited cards.
Raise 1X with a low pair.
Raise 1X with at least three points.
Raise 1X with a straight draw with no gaps, 4,5,6 or higher.
Raise 1X with a straight draw with one gap and two 1-point cards.
Fold all others.

4 Cards

Raise 3X with any made hand.

Raise 3X with any four to a flush.

Raise 3X with four to an outside straight, 8-high or better.

Raise 1X with any other straight draw.

Raise 1X with a low pair.

Raise 1X with at least four points.

Raise 1X with three 1-point cards and at least one previous 3X raise.

Fold all others.

10

Pai Gow

Pai gow is played with 32 tiles or dominoes. It should not be confused with Pai Gow Poker, which is based on pai gow, but uses cards instead of tiles. If you're looking for a game with a slow pace, low volatility, a challenging strategy, and Asian charm, pai gow may be the perfect game for you. In my experience, to know pai gow is to love pai gow.

Rules

Normally, I don't explain the rules of the games covered in this book, but pai gow is so little-known that it merits an exception.

- Pai gow is played with a set of 32 dominoes.

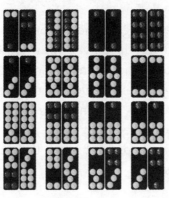

- The dealer and each player are given four tiles.
- The player separates his tiles into low and high hands of two tiles each. The player doesn't need to specify which is higher, as this is obvious.
- Each pair of tiles has a ranking order as follows:

    Pair: There are 16 pairs, as shown in the image on the previous page. The tiles are pictures in rank order, starting at the upper left and reading like a page to the bottom right.

    Wong: a 2 or 12 tile with any 9 tile.

    Gong: a 2 or 12 tile with any 8 tile.

    9 to 0 points (the more the better): For all other two-tile hands, the total number of dots is added together and the last digit is used to determine the number of points (as in baccarat). For example, a 10 and 9 tile (19) is worth 9 points; a 4 and 7 tile (11) is worth 1 point.

    An exception to the above rule is that the two tiles in the highest ranking "Gee Joon" pair are semi-wild and worth either 3 or 6 points, whichever results in a higher total. For example, when combined with a 4-point tile, a Gee Joon tile counts as 3 points to make a 7-point hand, instead of counting as 6 points to make a 0-point hand.

- The player's high hand is compared to the dealer's high hand and the player's low hand compared to the dealer's low hand (as in Pai Gow Poker).
- If both player and dealer have a Wong, Gong, or 1-9 points, the tie is broken according to which hand has the higher-ranked high tile.
- If the high tile doesn't break the tie, the win goes to the banker.

- A 0-0 tie always goes to the banker.
- When using high tiles to break a tie, the rank order is the same as the pair order (see image on pg. 45), except the two tiles in the highest Gee Joon pair are ranked lowest individually and are therefore never a hand's high tile.
- If the player wins both hands, he or she wins even money, less a 5% commission. If the player wins one and loses one, the wager is a push. If the dealer wins both, the player loses the wager.
- At most casinos, the turn to be banker rotates around the table. At some casinos, such as Foxwoods, the turn to be banker zig-zags between the player(s) and dealer, meaning that the dealer must bank at least every other hand.
- The player may invoke the turn to bank, although most players decline.
- The banker plays against every other player at the table and the dealer. The dealer wagers the same amount the player wagered the last time the dealer was the banker. More often than not, other players sit out the hand when another player is banking, letting him or her take on the dealer alone.

As you can see, the rules for pai gow are rather complicated. The strategy is even more difficult. It's the most difficult casino game to play and highly challenging to master. But the basic strategy I present later in this chapter is a good balance between simplicity and power.

House Edge

The house edge depends significantly on whether or not the player is banking and, to a lesser degree, on the player's strategy. The following table shows the house edge against three strategies whether banking or not.

TABLE 15—Pai Gow House Edge			
Strategy	Not Banking	Banking	Average
House Way	2.44%	0.53%	1.49%
Basic	1.98%	0.08%	1.03%
Optimal	1.66%	–0.20%	0.73%

What may jump out from the table is the 0.20% player advantage when banking and using optimal strategy. This, in theory, is the case on all pai gow games. It actually gets even better if you bank against multiple players, as the commission is charged against the net win, as opposed to each bet individually. However, the social dynamics at the table might make other players reluctant to bet against you. This thin advantage probably isn't worth pursuing as a professional gambler, but if you play pai gow anyway, it's better than the negative expected value of not banking.

Prepaying the Commission

Many casinos allow the player to "prepay" the commission—for example, betting $105 to win $100, as opposed to $100 to win $95. This effectively cuts the commission from 5% to 4.76%. Doing so reduces the house edge by 0.07%.

Banking an Extra 10%

Normally, when banking, the dealer bets against the player in the same amount the player bet against the dealer in the previous hand. However, some casinos allow the player to request that the dealer bet 10% more when banking. If the player is banking every other hand, this lowers the overall house edge by 0.07%.

Banking Against Other Players

If other players bet against you when it's your turn to bank, in theory, that's a good thing. The greater your percentage of the action with the banker advantage, the lower the combined "house" edge. However, unspoken etiquette at the pai gow table seems to be that only the biggest swinging banana at the table should dare invoke his right to bank. This is why I like to play alone, with friends, or with low-betting players who know their place. Asking to bank with an Asian tiger at the table probably won't go over too well.

Strategy

Although casino house ways can be more than 20 pages long, such strategies aren't very powerful. It's my understanding that a certain way of playing, which evolved over hundreds of years, has become more or less the house way. However, computer analysis has shown the house way to be much too conservative in terms of balancing the high and low hands. To maximize expected value, a more aggressive strategy is called for, often playing a strong high hand and a weak low hand, as opposed to balancing.

In the next section, I present what I call my "basic strategy" for pai gow. It was designed to emphasize simplicity, while minimizing the cost of errors, compared to optimal strategy. My object was to have it fit on a 4" x 5.5" laminated strategy card at a minimum of error cost.

If you want to go beyond the basic strategy, WizardOf Odds.com has more advanced strategies. These, of course, are more complicated. Optimal strategy is not only extremely difficult, but I doubt it's known completely by anyone on Earth. That said, a few people I know can play to near optimal levels.

The Basic Strategy

If two rules seem to contradict, go with the one listed first.

For purposes of this strategy, count hands of 9 points or less according to their number of points. Count Gongs as 10 points, Wongs as 11 points, and pairs as 12 points.

1. If there's one way to play a hand that's superior to both alternatives, play that way.

2. **Pairs**
    -   Never split 10s or 11s.
    -   If the point total by splitting the pair is greater than retaining the pair, then split it.
    -   If the point total by splitting or retaining the pair is the same and that total is less than 15, then split the pair.
    -   Otherwise, retain the pair.

3. **One High-9, Gong, or Wong Possible**
   - Play the best high.

4. **Two or Three Ways to Play High-9, Gong, or Wong**
   - With both the 2 and 12 tiles, always play the 12 in the high hand.
   - With 10 or 11 total points, maximize the high.
   - With 12 total points, maximize the high, except if it's a High-8 Gong, then balance.
   - With 13 total points, maximize the high, except if it's a High-8 Gong, then play a High-9.
   - With a point total of 14 or more, play High-9 if you can, otherwise Low-8 Gong and 5 in the low if you can, otherwise best low.

5. **All Other**
   - With a point total of 6 or less, balance.
   - With a point total of 7 to 9, play best high.
   - With a point total of 10 to 15, if you can get to 5 or more in the low, then balance. Otherwise, play best high.
   - With a point total of 16 or more, play best high.

A Unique Game

If every game in the casino had the same house edge, pai gow would be my favorite. Besides the challenge, I enjoy the Asian ambience. When I first starting playing, I was always the only pale-face at the game. When I sat down, dealers often asked, "Are you sure you're at the right table?"

After having written extensively about pai gow at Wizard OfOdds.com, I've introduced it to a wider audience. Today, I estimate about 15% of pai gow players in Las Vegas are non-Asian. When I sit down and play with non-Asians, I'm usually warmly greeted by someone who learned the game from my website and often has a copy of one of my strategies in his pocket (which I'm happy to autograph!). If I've played a role in integrating pai gow, I'm proud of it.

11

Pai Gow Poker

Pai Gow Poker is a good choice of game if you're looking for a "dribbler," meaning a game that offers a long time at the table with comparatively little risk. This happens by means of a slow rate of play and a high probability of either a win or push. The rules are fairly simple and it's easy to jump in and play, but mastering the strategy is quite difficult.

House Edge

The house edge depends significantly on whether or not the player is banking and, to a lesser degree, on the player's strategy. The following table shows the house edge against five strategies, whether banking or not. The Wizard Simple Strategy is provided later in this chapter. The intermediate and advanced strategies can be found at WizardOfOdds.com.

TABLE 16—Pai Gow Poker House Edge

Strategy	Banking	Not Banking
House Way	0.25%	2.72%
Wizard Simple Strategy	0.21%	2.67%
Wizard Intermediate Strategy	0.10%	2.57%
Wizard Advanced Strategy	0.05%	2.52%
Optimal Strategy	0.04%	2.51%

Much more important than strategy is that you exercise your option to bank as often as you can, as long as you're comfortable with fading the action the other players bet against you. Rules of when the player can bank vary from casino to casino and even dealer to dealer. It's good if the bank rotates around the table, as opposed to zig-zagging with the dealer. Better yet is when empty spots are skipped over, as in pai gow tiles, but this isn't often the case in Pai Gow Poker.

Simple Strategy

Much of the time, the right play is obvious in this game. If you're faced with an obscure situation and don't know what to do, asking how the dealer would set the hand, known as the "house way," is always a safe choice. Although the house way isn't always right, it usually is, and when it isn't, the degree of error is generally small.

However, for aficionados of the game, I present my Wizard Simple Strategy. It's divided first by whether or not a straight or flush can be played, then by the highest hand that can be made, other than a straight or flush, and finally by other specifics about that high hand and the other cards not part of it. This strategy is optimized against the Foxwoods house way, where A-2-3-4-5 is the second highest straight, but all house ways are more or less the same.

Simple Strategy—No Straight or Flush Possible		
5 Aces		AA in low
4-of-a-Kind	+ 3-of-a-Kind	Play best low
	+ 1 Pair	Play pair in low
	2,3,4	Play highest two singletons in low
	5	Play 10-high or better in low, otherwise split apart
	6	Play Q-high or better in low, otherwise split apart
	7,8	Play K-high or better in low, otherwise split apart
	9,10,J	Play A-high or better in low, otherwise split apart
	Q,K,A	Split apart
3-of-a-Kind	+ 3-of-a-Kind	Play best low
	+ 2 Pair	Play higher pair in low
	+ 1 Pair	Play pair in low
	A	Play ace and highest singleton in low
	All other	Play two highest singletons in low
3 Pair		Play highest pair in low
2 Pair*	6 or less	Keep together with Q-high or better in low, otherwise split apart
	7 to 11	Keep together with K-high or better in low, otherwise split apart
	12 to 16	Keep together with A-high or better in low, otherwise split apart
	17 or more	Always split apart
1 Pair		Play best low hand
No Pair		Play best low hand

*Here are the points awarded to each pair: A = 14; K = 13; Q = 12; J = 10; 2-10 = pip value. Add the following point value of each pair and then use the strategy to see when to retain the two pair as opposed to splitting apart.

Simple Strategy—Straight or Flush Possible		
4-of-a-Kind		Play AA in low
3-of-a-Kind	+ 1 Pair	Retain the straight or flush if you can play 77 or better in low, otherwise play the pair in low
	All other	Play best low while maintaining straight or flush in high
3 Pair		Play AA in low
2 Pair*	6 or less	Keep together with Q-high or better in low, otherwise split apart
	7 to 11	Keep together with K-high or better in low, otherwise split apart
	12 to 17	Keep together with A-high or better in low, otherwise split apart
	18 or more	Always split apart
1 Pair	9 to Q	Play AK if possible in low and pair in high, if possible, otherwise play best low while maintaining straight or flush in high
	All other	Play best low while maintaining straight or flush in high
No Pair		Play best low while maintaining straight or flush in high

*Here are the points awarded to each pair: A = 14; K = 13; Q = 12; J = 10; 2-10 = Pip value. Add the following point value of each pair and then use the strategy to see when to retain the two pair as opposed to splitting apart.

Side Bets

Pai Gow Poker has a host of side bets. I could easily recommend that you just avoid all of them in every game, but here are a few common ones, with their paytables and house advantage.

Fortune—This one has many paytables available. The

most common seems to be 2-3-4-5-25-50-150-400-1000-2000-8000. The house edge is 7.77%, less 0.93% for every additional player at the table (besides yourself), due to the Envy Bonus.

Emperor's Challenge—The only paytable I'm aware of is 40-5-2-2-3-4-5-25-50-150-500-1000-5000, when ranking hands from low to high, and has a house edge of 4.17%.

Pai Gow Insurance—The paytable 3-5-7-15-25-100 has a house edge of 7.35%.

Variants

A few Pai Gow Poker variants are commission-free. However, to pay for that, the gamemakers quietly change other rules that favor the dealer. One rule common to them all is no player banking. Here are some of the commission-free pai gow poker games and other rules changes, to allow for removal of the commission.

EZ Pai Gow Poker—This variant, invented by the late Dan Lubin, a good friend of mine, automatically pushes all hands if the dealer's best hand is a queen-high. The house edge is 2.47%. A host of side bets may be tacked on, which I won't get into.

Commission-Free Pai Gow Poker—This variant automatically pushes if the dealer plays a 9-high in the low hand. The house edge is 2.51%. The Tiger 9 bet pays 30-1 if the dealer plays a 9-high in the low at a house edge of 22.48%.

Face-Up Pai Gow Poker—This variant not only gets rid of the 5% commission, but the dealer's cards are dealt face up. To pay for that, all dealer ace-high hands are an automatic push. The house edge is 1.81%. The Push Ace High side bet, with a 5-15-40 paytable, has a house edge of 9.29%.

12

Roulette

Roulette is one of the easiest casino games to analyze. The only strategy is in game selection and, sometimes, bet selection. Only two rules affect the odds of the game:

- Number of zeros. This can be one, two, or three.
- What happens on even-money bets if the ball lands on a zero. Either they lose half or lose all.

The primary purpose of this chapter is to address the house edge of the five major known variations of the rules — from best to worst.

French Roulette

This term refers to roulette played with a single zero where the player gets back half the bet if the ball lands on zero. The name French roulette is somewhat of a misnomer, because this set of rules can be found outside of France and what you find in France is probably single-zero roulette.

Whatever it's called, the house edge on even-money bets is as good as it gets in roulette, 1.35%. On all other bets, it's 2.70%.

I should comment that some older gambling books mention an "en prison" rule, where if the ball lands on zero, even-money bets are covered in a cage, representing a prison, and the next spin determines whether the bet loses or is a push. There are different accounts of what happens on a second consecutive zero. Mathematically, losing half and the en prison rule are very similar. Suffice it to say that I've been to Europe six times and have never actually seen the en prison rule practiced. I believe it to be an archaic rule we can stop talking about.

Single-Zero Roulette

This is also called European roulette, especially by Internet casinos. It's the same as French roulette, except even-money bets lose all if the ball lands on zero. Thus, the house edge is 1/37, or 2.70%, on all bets.

Atlantic City Roulette

This is the same as French roulette, except there are two zeros instead of one. This results in a house edge of 2.63% on even-money bets and 5.26% on all others, except the 0-00-1-2-3 combination ("five-number bet"), which comes in at 7.89%.

Double-Zero Roulette

This is the same as single-zero roulette, where even-money bets lose all on a zero, except there are two zeros. The house edge is 5.26% on every bet, except the dreaded 0-00-1-2-3 combination at 7.89%.

Triple-Zero Roulette

This is also called Sands roulette. It's the same as double-zero roulette, except with three zeros. It started at the

Venetian around 2015 and has since spread to other properties in Las Vegas. This game seems to get equal or better play
than double-zero roulette at the same minimum bet. Don't
ask me why. The house edge on every bet is 7.69%.

Side Bets

Yes, side bets have reached roulette too. I won't waste
space describing all of them, because none of them is particularly widespread. For all of them, the house edge on the side
bet is greater than the base game. If you want to score a big
win, just bet on a single number.

Dealer Signature

I've heard many times that some dealers are so delicate,
they can steer the ball to land in a desired sector of the wheel.
The amount of evidence I've seen to support this, other than
anecdotal stories, is zero. Much like dice influence, I'm very
skeptical of it (see "Gambling FAQ").

Biased Wheels

If a wheel isn't balanced, it will favor a certain sector over
the one opposite of it. This can take a sampling of thousands
of spins to tease out random variation and detect a correlation.
My opinion is wheels in the United States are very high-qual-
ity and well-maintained, so any minuscule bias wouldn't
overcome the house edge. However, in less affluent countries,
the wheels may not be up to our standards, so I believe some
advantage players still exploit them. Stories have come out

in the past about European teams beating biased wheels for large amounts of money, but I imagine it's probably more difficult to get away with than it used to be. That's usually the story with advantage play: By the time you hear about a play, it isn't good any longer.

13

Slots

Between a high house edge and a fast rate of play, I believe slot machines to be the quickest way to lose your money in the casinos. To anyone reading this who plays slots, I highly recommend that you convert to video poker. The only benefit to playing slot machines is that you get comped well, but that's only because slot players lose so much. Even with the generous comps, you'll still lose much more than the casinos give you back.

House Edge

The odds on slot machines are very difficult to quantify. In England, they kindly (actually, they're mandated to) disclose the return percentage on each machine, with returns generally in the low 90% range, as I recall from a trip to London in 2014. However, here in the colonies, I know of no such consumer-protection laws. Slot players simply have no right to know the odds of the machine they're playing, other than minimum return percentages of about 75% that exist in some states, including Nevada.

While we don't know the odds on a specific machine, we can make generalizations about slots. In every non-tribal gaming jurisdiction in the United States that I'm aware of,

some form of gaming-revenue reporting is required. Sometimes there's a line item for slot revenue and sometimes it's
mixed into general casino revenue. In late 2018, I searched
through what data I could find and found the following states
that report slot revenue separately.

TABLE 17—Slot Revenue by State

State	Period	Revenue	Handle	House Edge
Colorado	2017	$ 722.46	$ 9,864.84	7.32%
Indiana	2017	$1,908.00	$ 20,578.70	9.27%
Iowa	2016	$ 358.32	$ 3,923.54	9.13%
Maryland	2016	$ 358.32	$ 6,893.26	5.20%
Mississippi	2017	$1,735.72	$ 22,447.66	7.73%
Nevada	2017	$7,283.08	$108,379.21	6.72%
Pennsylvania	2017	$2,336.19	$ 29,961.93	7.80%

Note: Revenue and handle are in millions.

I should note that the way casinos use the word "slots" is
to mean any game that doesn't require a human dealer. This
would include video poker, video keno, and electronic table
games. However, reel slots make up the lion's share of that
sector. Video poker and electronic table games have much
better odds, so if we were to tease those out, we'd find that
reel slots have an even higher house advantage than the table
above suggests. The odds in video keno I find to be about the
same as simple reel slots.

Advice

Following is some advice, if you absolutely must play the slots.

- Always use a players card.
- The simpler the machine, the better. Elaborate machines based on popular movies and television shows are set to a house edge of about 15%. Simple games without any name branding have a house edge of about 6%-12%, depending on many factors.
- In Las Vegas and Reno, as a rule of thumb, the fancier the casino, the tighter the slots. The locals casinos away from the Strip and downtown generally offer better odds. My source is the 2002 survey I conducted of every casino in Las Vegas.
- Never play slots, or anything, at the Las Vegas airport. Everything (slots, video poker, and video keno) is set to the stingiest possible rules the game manufacturers offer, which doesn't surprise me given the airport's captive and transient audience.

Advantage Plays

Some slot machines have "must-hit-by" jackpots, for example, a jackpot that must hit by $500. If you play these games only when they're very close to the maximum jackpot, the odds should be in your favor. However, I'd like to warn you that games by AGS design the mechanism that determines when the jackpot is hit to usually delay it until it's near the must-hit-by point. For example, in a game called River Dragons, the small jackpot starts at $200 and must hit by $500, while the large jackpot starts at $4,000 and must hit by $5,000. Normally, you'd expect the jackpot to hit on average halfway between these two markers—$350 on the small and $4,500 on the large. However, on this River Dragons game, I find that the small jackpot hits on average at $490.39 and the large jackpot at $4,945.49.

Some machines can be in different phases of the game, depending on previous spins. As a generality, I say that the odds are the same for every spin, but there are exceptions. An old example of an exception is the game Green Stamps, where players earned stamps as they played and when they accumulated 1,200, it was good for five free spins. If the previous player abandoned the game with 1,199 stamps it was an ideal time to jump in and "vulture" that bonus (to use an advantage player term).

A more modern example is Ocean Magic. In this game, wild symbols aren't positions on the reel strips; instead, they bubble up randomly from the bottom of the screen and land on the bottom row, replacing some other symbol that would naturally be there. Over the next three spins, the wild bubble floats up one position on the screen (there are four visible positions on each reel). So if the previous player leaves a bubble on one of the bottom three rows, preferably on the left side, the next player has a nice advantage on the subsequent spins while the bubble remains on the screen. As soon as it floats off the top row, it's a good time to get up and leave. My apologies to the advantage players who already know this one.

Skill-based slots also provide a possible advantage for "skilled" players. These games have a skill-based bonus. The prize pools available in these bonuses depend in part on the skill of the *previous* players. If they didn't do well at the bonus, it leaves more in the bonus bank for you to win. Some sort of meter on the game should show the maximum win and what you have to do to get it. In my limited experience, I find the skill level required in these games to be pretty low, so there usually isn't enough money in the bonus bank to be worth playing. However, with a lot of patience and time, you can find some ripe games.

In any form of gambling, comps can be a function of play and/or how much you lose. There are various ways of tricking casinos into thinking you lost more than you really did, some of which could be considered cheating, at least ethically, depending on the reason and method. With slots, for example, if a meter is nearly full and you know you're near a point where a bonus is triggered, you can pull out your players card until you trigger and play the bonus. This way, you'll be playing the bonus off the radar and it will look to the marketing department that you lost more or won less than you really did.

Be advised that this practice, known in the industry as "card pulling," is, to put it mildly, frowned upon by the casinos. Should you be caught (or even suspected of doing it), there will likely be consequences, ranging from being "no-mailed" to outright exclusion from further play.

Final Thought

I have two words (and an exclamation point) for the recreational slot players out there: Don't play! They offer terrible odds and very short times between bets, and don't have the decency to tell you the odds against you. If you must have spinning reels to be happy, plenty of apps and websites offer slot games for free.

14

Sports Betting

This chapter endeavors to impart some basic sports betting advice for the non-handicapper. My general advice can be summarized in three words: underdogs, visitors, and unders. However, the strength of this advice varies from sport to sport, as we shall soon see.

Entire books have been written on sports betting and they barely scratch the surface of this complex and mathematically precise gambling option, which is currently exploding throughout the nation. I've crunched a lot of numbers to show you how you can place bets that give you the best chance, in terms of recent history, of cashing your tickets. Keep in mind that these are rules of thumb and you could spend a lifetime perfecting methods for getting an edge on the sports books — and plenty of wiseguys and smart money do just that. Here, we're keeping it simple for recreational bettors who just want a nice inexpensive bang for their gambling buck. Making an $11 wager on a two- to three-hour sporting event (to win $10, for a $21 total return) is among the greatest bargains in the betting universe, especially when you factor in the extra excitement of rooting for your money.

First, I focus on the basic "off-the-board" bets (live-betting lines) for the NFL, MLB, and NBA. Then, I offer some general advice that applies to multiple sports.

In all tables, win rates are based on bets resolved (in other words, not counting pushes). Expected returns on money-line and run-line bets are based on flat betting one unit per game. In all sports, win probabilities and expected values are based on between 3,000 and 7,000 games, a relatively small sample size, so please take any results with a grain of salt.

NFL

The following NFL tables are based on data from 3,220 pro-football games played from the 2006 season through week 1 of the 2018-2019 season.

Let's start by looking at bets against the pointspread, laying 11 to win 10 (which requires a win rate above 52.4% to be profitable). I cut the data a number of ways, as Table 18 shows. You can see that flat betting road teams has an expected value of positive 1.34% against the spread ("ATS"). Better yet, betting road underdogs is +2.57% ATS. I'm not claiming these bets actually have a theoretical player advantage. Rather, I believe they've exceeded expectations in the limited sample. I do strongly believe, however, that they're better than the alternative of betting favorites and home teams.

Why have road teams fared better? The old adage is that the home-field advantage is worth 3 points in the NFL. However, the data show home teams outscore road teams by only 2.4 points on average. Furthermore, based on anecdotal evidence only, square bettors simply love to bet a strong team at home and will lay whatever points necessary to do so. This demand creates value betting the opposite way.

TABLE 18—Results ATS (NFL)

Bet	Win	Loss	Push	Win Rate	Expected Value
All home	1484	1680	56	46.90%	−10.28%
All road	1680	1484	56	53.10%	1.34%
All favorite	1488	1617	56	47.92%	−8.36%
All underdog	1617	1488	56	52.08%	-0.57%
Home underdog	492	520	20	48.62%	−7.05%
Home favorite	968	1125	36	46.25%	−11.51%
Road underdog	1125	968	36	53.75%	2.57%
Road favorite	520	492	20	51.38%	−1.87%

Next, let's look at betting the over/under line. The following table shows the results both ways. As you can see, there's little difference between the two. What little difference that exists is statistically insignificant. This does surprise me somewhat, as I was expecting under bets to do better, based on older data.

TABLE 19—Results Over/Under Bets ATS (NFL)

Bet	Wins	Loss	Push	Win Rate	Expected Value
Over	1,586	1,576	58	50.16%	−4.17%
Under	1,576	1,586	58	49.84%	−4.76%

The next table shows the results of money-line ("ML") bets. The greater expected value on underdogs does not surprise me.

TABLE 20—Results ML Bets (NFL)	
Bet	**Expected Value**
Underdog	–3.91%
Favorite	–6.24%
All	**–5.08%**

In general, you need to get more than 52.38% of your picks correct to beat laying 110 juice, not counting pushes. This is very difficult to do long-term.

MLB

The following MLB tables are based on data from 4,929 games played during the 2016 and 2017 seasons. (I had to omit two games due to incomplete information.)

The best type of bet in baseball is a 10¢ money line. By "10¢ lines," I mean there are 10 basis points between the two teams, for example:

Baltimore Orioles –140
Cleveland Indians +130

Many sports books have 15¢ or 20¢ lines, so it pays to shop around. The narrower the spread, the better for the player. Also, some sports books begin to widen the gap between the two lines earlier at lower numbers than others, so be mindful of that.

The expected value of bets on road teams on the money line is slightly better than home teams. To be specific, the expected value is 0.31% higher on road teams.

I show a greater disparity between underdogs and favorites. To be specific, underdogs have a 0.70% higher expected return on the money line than favorites.

Over/under bets show a strong preference toward the under. The following table shows the results against the usual 20¢ line.

TABLE 21—Results Over/Under Bets ATS (MLB)

Bet	Wins	Losses	Push	Win Rate	Expected Value
Over	2315	2373	241	49.38%	–5.45%
Under	2373	2315	241	50.62%	–3.20%

The run line works like a pointspread. Most sports books consistently offer a run line of –1.5 on the favorite, which is huge in baseball, as many games end in a margin of victory of 1. The expected return is greater laying the 1.5 runs on the favorite than getting them on the underdog. To be specific, the data show an expected return of –3.10% giving the 1.5 runs, compared to –3.66% taking them, a difference of 0.56%.

NBA

The following NBA tables are based on data from 6,567 games played from 2013 through 2017. The first table shows the results of road and home teams betting against the spread, laying 11 to win 10. As you can see, the road teams perform significantly better.

TABLE 22—Results Home/Road Bets ATS (NBA)					
Bet	Wins	Loss	Push	Win Rate	Expected Value
Road	3280	3159	128	50.94%	–2.70%
Home	3159	3280	128	49.06%	–6.22%

The next table shows the results betting favorites and underdogs against the spread. Not surprisingly, underdogs did better, but the disparity is only 0.77% in expected value, less than I was expecting.

TABLE 23—Results Favorite/Underdogs Bets ATS (NBA)					
Side	Win	Loss	Push	Probability	EV
Favorite	3176	3202	128	49.80%	–4.84%
Underdog	3202	3176	128	50.20%	–4.07%

Against the over/under, as expected, bets on the under did better than bets on the over, but not significantly. The following table shows bets on the under had an expected value 0.45% higher.

TABLE 24—Results Over/Under Bets ATS (MLB)					
Side	Win	Loss	Push	Probability	EV
Over	3232	3246	89	49.89%	–5.16%
Under	3246	3232	89	50.11%	–4.71%

Against the money line, road teams had an expected value of –3.71% and home teams –4.24%. Cutting it by underdogs and favorites, underdogs had an expected value of –4.21% and favorites –3.75%. This higher return of the favorites truly surprised me, especially given that underdogs did better against the spread. With only 6,567 games in the sample size, I think random variation is probably the reason.

By the way, the home team won 58.4% of the time and the average margin of victory was 2.70 points in favor of the home team.

As I mentioned at the beginning of this chapter, as a rule of thumb, I recommend you make the type of simple bets already discussed and avoid the exotic bets. However, there are many exceptions to this rule. The following sections give you some very general advice on exotics that apply to all sports.

Parlays

If you make a parlay bet off the board, mix in at least one bet that isn't laying 11 to win 10. If at least one bet on your parlay doesn't pay –110 by itself, it forces a mathematical calculation as if you bet each leg individually and let the winnings ride each time. The odds when everything pays –110 are stingy by comparison, with some sports books stingier than others. An exception to this is the 3-team parlay that pays 6-1, which is a little better than the 5.96 you'd get with a direct calculation.

Here's an example. Suppose you want to make a five-team parlay. If every pick is –110, then most sports books will pay 20-to-1 if all of them win. However, if you change one pick to a side that pays –115, then you'll go off the parlay pay-table and force a calculation. In that situation, the bet would pay $(210/110)^4 * (215/115) - 1 = 23.83$ to 1. Not only are you getting better odds than the 20, but the side that pays –115 should have a slightly better than 50% chance of winning.

I would generally avoid parlay cards, but if you do bet them, look for "stale" lines that have moved in your favor between when the card was printed and the time you fill it in. It's especially valuable to get a line movement of 3 points in football. On half-point parlay cards, bet only when you're getting a half-point off a flat spread, preferably 3, or a significant line movement. Be warned that if the line movement is too great, the sports book will probably "circle" the game, meaning that any parlay cards with that pick will not be accepted.

Teasers

With NFL teasers, whether off the board or on a teaser card, I would pick a team only if you're crossing both the critical 3- and 7-point margins of victory.

A couple of examples are teasing an underdog from +2 to +8, or a favorite from –7.5 to –1.5. These "Wong teasers" used to be a good advantage play, but every sports book has depressed its odds to the point where even these are no longer viable. With teaser cards, much like parlay cards, try to take advantage of stale lines.

Overall, however, you shouldn't waste your time and money on teasers. This advice goes for pleasers as well. There are some rare exceptions, but I won't list them, because this is a book and the information is likely to get outdated.

Proposition Bets

Back in the day, I loved Super Bowl proposition bets. They used to be a pretty reliable way to make about a 10% return on your money. However, this secret has long since been out of the bag (perhaps, in part, because of me). Nowadays, you have to compete with more sharp prop bettors with deep pockets who push lines closer to fair. Nevertheless, I think there is probably small value in betting NFL proposition bets on unders and anything *not* to happen (by "not to happen," I mean the "No" in Yes/No propositions, e.g., "Saquon Barkley will score a TD"). It especially pays to shop around, as prices vary significantly on props.

Shopping

It's difficult to overstate the importance of shopping around in any type of sports betting. Every sports book family sets its own odds and some may be out of balance on a game, causing them to incentivize the side they need with better odds than the sports book next door. Several websites, e.g., GamblingWithAnEdge.com, post up-to-date odds on major games at many Las Vegas and offshore sports books, which allows comparison.

As an example, as I write this, the line on the Tennessee Titans against the Philadelphia Eagles is +3.5 at the MGM sports books and +4 at William Hill. It may not sound like much of a difference, but 5.2% of NFL games end in a margin of victory of four points. Granted, it will only matter if the Titans win by four, but it's good to capture any free point you can, and four is the fourth best (after 3, 7, and 10).

Strategy

Here are my rules of thumb that apply to all sports.

- Underdogs over favorites
- Road teams over home teams

- Unders over overs
- Given a choice, look for tight money-line spreads between the two sides of a bet.
- Avoid cards and exotic bets.
- *Shop around.*

15

Texas Hold 'Em Bonus

Texas Hold 'em Bonus has been around since at least 2005 and continues to be one of the more popular poker-based games. It's a player-versus-dealer game based on hold 'em with three opportunities to raise as the cards are revealed. The good news is that with optimal strategy and the liberal Las Vegas rules, the element of risk is a low 0.53%.

Analysis

The base game involves an Ante bet and up to three subsequent raise bets, which are all interconnected. To evaluate the value of the overall game, you have to consider the value of each bet, which is shown in the following table. This table is based on the liberal Las Vegas rules where the Ante pays on a winning straight or higher.

TABLE 25—Texas Hold 'Em Bonus House Edge

Bet	Average Wager	Return
Ante	1.000000	–0.411959
First raise	1.909502	0.029522
Second raise	0.445910	0.163924
Third raise	0.462714	0.198144
Total	**3.818126**	**–0.020369**

The lower right cell in the preceding table shows that the house edge, as defined as the ratio of the expected loss to the Ante bet, is 2.04%. However, the middle column shows that by the end of the hand, you'll have bet 3.82 units. This makes the element of risk 2.04%/3.82 = 0.53%. Few games get that low.

Atlantic City Variant

In Atlantic City and perhaps other locations, they follow a stingy rule where the player needs a winning flush or higher to win the Ante bet, as opposed to the Las Vegas rule, where only a straight is required. The Atlantic City rule increases the house edge from 2.04% to 5.59%.

Strategy

I have quantified the strategy of this game only for the initial decision point. That strategy is pretty simple, as follows:

- *Las Vegas rules:* Raise on anything except unsuited 2-3 to 2-7.
- *Atlantic City rules:* Same as Las Vegas strategy, except also fold suited 3-4.

While I haven't quantified the strategy past the first decision point, I can say that under the Las Vegas rules, the player who makes the Flop bet will make the Turn bet 43.13% of the time. If the player makes the Turn bet, he'll also make the River bet 85.76% of the time. If the player makes the Flop bet, but not the Turn bet, he'll make the River bet 15.78% of the time.

Bonus Bet

The one side bet that always seems to accompany this game is the Bonus. I've seen two variants of it. Both start with a paytable of 3-5-10-15-20-25-30. If 30 is the highest possible win (for player pocket aces), the house edge is 8.90%. Sometimes an extra level is added that pays 1,000 if both the player and dealer have pocket aces. This lowers the house edge to 8.54%.

Other Side Bets

You'll sometimes see the multi-game 6-Card Bonus. This side bet is addressed in the chapter on Crazy 4 Poker. Casino Bregenz, in Austria, calls this game Easy Poker and adds a side bet called the C5, which carries a house edge of 8.95%.

If you see any side bets not mentioned above, just remember my Ninth Commandment of gambling: "Thou shalt not make side bets."

16

Three Card Poker

Three Card Poker is one of the earliest and most successful of a wave of privately owned casino games. It was invented by Derek Webb in 1994 and continues to attract players looking for an easy-to-play poker variation.

Ante

The odds (and element of risk) on the Ante bet depend on the paytable used for the Ante Bonus. Four different paytables are represented in the following chart. The most commonly seen Ante Bonus paytable is the original 1-4-5. A frequently made dealer error is to not pay the Ante Bonus when the dealer beats the player hand. The Ante Bonus is supposed to pay regardless of the dealer hand.

TABLE 26—Ante Bet House Edge				
Player Hand	**Table 1**	**Table 2**	**Table 3**	**Table 4**
Straight Flush	5	4	3	5
3-of-a-Kind	4	3	2	3
Straight	1	1	1	1
House Edge	3.37%	3.83%	4.28%	3.61%
Element of Risk	2.01%	2.28%	2.56%	2.16%

Pair Plus

When Three Card Poker first appeared, it offered the liberal 1-4-6-30-40 paytable for several years. Gradually, stingier paytables replaced it. The most common one today is the same as the original, except the pay on the flush was lowered from 4 to 3. This may not seem significant, but it increases the house edge from 2.32% to 7.28%. The following table indicates the house edge for eight known paytables.

Player								
TABLE 27—Pair Plus Bet House Edge								
Player Hand	**Paytable**							
	1	**2**	**3**	**4**	**5**	**6**	**7**	**8**
Straight Flush	40	35	40	35	50	40	40	40
3-of-a-Kind	30	33	25	25	30	30	25	30
Straight	6	6	6	6	6	5	5	6
Flush	4	4	4	4	3	4	4	3
Pair	1	1	1	1	1	1	1	1
Nothing	−1	−1	−1	−1	−1	−1	−1	−1
House Edge	2.32%	2.70%	3.49%	4.58%	5.10%	5.57%	6.75%	7.28%

Some paytables have a line item for a Mini Royal, which is a suited Q-K-A. For every 10 the Mini Royal pays above a straight flush, the house edge on the Pair Plus is lowered by 0.18%. For example, a common paytable that pays on a Mini Royal looks like this: 1-3-6-30-40-50. The Mini Royal pays 50, which is 10 more than the pay for a straight flush, thus the house edge is lowered from 7.28% to 7.10%.

Other Side Bets

Through the years, many other side bets have been added to Three Card Poker. I can confidently say that they're all sucker bets to be avoided. Nevertheless, I'll briefly go over some of them.

The Three Card Progressive is a $1 "red-light" bet (the betting spot is a red light on the table) that pays a progressive jackpot for a Mini Royal in spades. It also pays an Envy bonus if another player gets a Mini Royal in any suit. The paytable 90-100-500 has a return of 47.87%, plus 4.52% for each $1,000 on the meter and 0.79% for each player besides yourself. The paytable 6-60-70-500 returns 54.39% plus the same additions for the jackpot and other players.

The Five Card Progressive is also a $1 red-light bet, but pays based on the player's three cards and two other cards. It pays the full jackpot for a royal flush and 10% for a straight flush. The rest of the paytable is 9-30-40-50. Envy bonuses are $1,000 for a royal flush and $300 for a straight flush. The return is 53.06% plus 2.92% for each $10,000 in the meter and 0.57% for every other player at the table.

The Prime pays 3-1 if the player's three cards are the same color and 4-1 if the dealer's cards are also of that same color. The house edge is 3.62%.

The Six Card Bonus pays based on the poker value of the best 5-card hand that can be composed of the six cards between the player and dealer hands. Following are three paytables I've seen and their associated house edge:

7-10-15-20-100-200-1000	8.56%
5-10-15-25-50-200-2000	14.36%
5-10-15-25-50-200-1000	15.28%

The Millionaire Maker is the same as the 6-card-bonus, except it adds wins of $100,000 for a "Super 6 Card Royal Flush" and $1,000,000 if it's in diamonds, as long as the bet is at least $5. The 5-10-15-20-50-200-1000 paytable for the other hands has a house edge of 18.10%.

Strategy

The strategy is very simple: Raise with Q,6,4 or better.

Hole-Carding

Sometimes, when the dealer takes his hand out of the shuffler, he will flash the bottom card to the table. This used to be a great advantage play. Unfortunately, dealing procedures have been mostly changed to guard against it, but some dealers are still flashing out there. Here is the strategy if the player can clearly make out the rank of one of the dealer's cards, based on the card exposed.

- 2-J: Raise always.
- Q: Raise with Q,9,2 or better.
- K: Raise with K,9,2 or better.
- A: Raise with A,9,2 or better.

Assuming the 1-4-5 Ante Bonus paytable, the player advantage against a flashing dealer using this strategy is 3.48%.

17

Ultimate Texas Hold 'Em

Ultimate Texas Hold 'em is probably the most popular new table game since Three Card Poker. The game features the excitement of large raises, big wins for premium hands, and a challenging strategy. Most players don't realize it also has some of the best odds of any casino game, if played properly.

In fact, Ultimate Texas Hold 'em deserves a place on the short list of the best casino games in terms of player value. The only other easily found games that might offer better odds, depending on the rules, are blackjack, craps, and video poker.

Not only are the odds highly competitive, the game is fun and exciting to play. But I suspect the real reason you see so many Ultimate Texas Hold 'em tables is simple—player errors. Few players have any clue about proper strategy.

Ante, Blind, and Play Bets

The Ante, Blind, and Play bets are all connected to one another and are essential to the game. The following table shows the average amount of each wager, based on a one-unit Ante bet and the expected return for each, under optimal strategy.

TABLE 28—Ultimate Texas Hold 'Em House Edge		
Bet	**Average Wager**	**Expected Return**
Ante	1	–0.165757
Blind	1	–0.314685
Play	2.152252	0.458593
Total	**4.152252**	**–0.021850**

What the table above tells us is that by the end of the hand, the player can expect to lose an amount equal to 2.185% of his Ante bet. That may sound average for a poker variant, but I think that house-edge statistic is misleading in regard to how good a bet Ultimate Texas Hold 'em is. For comparing one game to another, I prefer the element of risk. With a required two-unit starting bet and an average Play bet equal to 2.15 units, the player will bet an average of 4.15 units by the end of his hand. That makes the EOR a very competitive 2.185%/4.15 = 0.53%.

Optimal strategy in Ultimate Texas Hold 'em would be extremely long and tedious to memorize. The number of people who know it perfectly is probably zero. The good news is you can get close to optimal with my basic strategy below. Using it results in an EOR of 0.58%, only 0.05% higher than optimal strategy.

Large Raise

The large raise strategy is easy to quantify perfectly.

- Make the large raise with any pair except deuces.
- With two singletons, make the large raise according to the following table that shows the lowest hand in each category for which the odds favor a large raise. For

example, with an unsuited king-high, you'd make the large raise with K,5 or better. Never make the large raise with 10-high or less.

High Card	Suited	Unsuited
Ace	Any	Any
King	Any	K-5
Queen	Q-6	Q-8
Jack	J-8	J-10

My strategy for the medium and small raises consists of some rules of thumb that fit most situations. These rules were inspired by the James Grosjean strategy card, which I highly recommend if you're looking to get even closer to optimal strategy at a cost of a little more to remember.

Medium Raise
Make the medium raise with any of the following:

- Two pair or better.
- Hidden pair*, except pocket deuces.
- Four to a flush, including a hidden 10 or better, to that flush

Small Raise
Make the small raise with any of the following, otherwise fold:

- Hidden pair or better.
- Fewer than 21 dealer outs that beat you.

*Hidden pair = any pair with at least one card in your hole cards.

What's a "dealer out," you might ask? It means a dealer's hole card that will beat you.

Let's look at an example. Suppose the board is K♥ 7♠ 2♥ A♣ 10♥ and you have a jack-high. There are 15 cards that will pair the dealer and beat you (three kings, three 7s, etc.). Any one of the four queens will also beat you. That makes 15+4 = 19 cards that will immediately beat you. Since 19 is less than 21, the odds favor making the small raise.

Never mind any two-card combinations that can also beat you. That just muddies the water. So you can ignore the fact that the dealer has three to a flush in this example.

Trips Bet

Trips is a side bet that pays based on the poker value of the player's hand only. I'm aware of four paytables, as follows:

TABLE 29—Trips Side Bet House Edge				
Player Hand	**Paytable**			
	1	**2**	**3**	**4**
Royal Flush	50	50	50	50
Straight Flush	40	40	40	40
4-of-a-Kind	30	30	30	20
Full House	9	8	8	7
Flush	7	6	7	6
Straight	4	5	4	5
3-of-a-Kind	3	3	3	3
House Edge	**0.90%**	**1.90%**	**3.50%**	**6.18%**

Paytable 3 seems to be the most common. While a 3.5% house edge isn't bad for a side bet, it's still much worse than the base game.

Progressive

I've seen three different progressives on Ultimate Texas Hold 'em, but the most popular at the time of this writing is the one with the following payouts. This red-light bet is always for $1 and wins are on a "for-1" basis.

Royal Flush	100% of jackpot
Straight Flush	10% of jackpot
4-of-a-Kind	$300
Full House	$50
Flush	$40
Straight	$30
3-of-a-Kind	$9

The return percentage is 53.06%, plus 2.92% for every $10,000 in the meter. The breakeven point at which the return is 100% (meaning a zero house edge), is when the meter is at $160,530.53.

18

Video Keno

Live keno is going the way of the dodo bird, but video keno is alive and well. In addition to conventional video keno, known as "spot keno" in the industry, several variants also appear on casino floors around the country. This chapter provides the return by the number of picks and paytables for the most common video keno games. The paytables are taken from Game King machines, which dominate video keno.

To use these tables, look up the sequence of wins for any game and number of picks in the appropriate table to find the associated return. For example, consider the following paytable for a 5-spot:

Catch	Pays
0	0
1	0
2	0
3	3
4	13
5	838

This paytable would be abbreviated as 3,13,838. From the Pick 5 paytable for spot keno, we see it has a return of 94.95%. Note that, in the interest of making efficient use of space, leading zeros have been removed from the tables. In the previous example, the full paytable would be denoted 0,0,0,3,13,838, but I've dropped all the zeros that pay nothing. I use an asterisk if catching zero pays 1.

Spot Keno

This is the conventional keno game, where the player picks 2 to 10 numbers, the game picks 20, and the player wins according to the number of picks he's made that match the draw by the game.

Pick 2

Paytable	Return
14	84.18%
5*	86.08%
15	90.19%

Pick 3

Paytable	Return
1,2,11	84.66%
2,40	83.25%
2,43	87.41%
2,45	90.19%
2,46	91.58%
3,37	92.96%
2,48	94.35%

*Catching zero pays 1.

Pick 4

Paytable	Return
1,1,5,40	85.97%
2,3,100	86.14%
2,5,77	87.74%
2,5,85	90.19%
2,5,91	92.03%
1,12,64	92.77%
2,5,100	94.78%

Pick 5

Paytable	Return
2,14,800	85.31%
1,3,10,400	85.79%
3,12,750	88.06%
3,11,804	90.33%
3,12,810	91.93%
1,3,28,170	92.72%
3,13,838	94.95%

Pick 6

Paytable	Return
1,2,5,49,1000	84.96%
2,4,92,1500	85.21%
3,4,55,1600	88.02%
3,4,68,1500	90.76%
3,4,70,1600	92.67%
1,1,14,62,300	92.66%
3,4,75,1660	94.99%

Pick 7

Paytable	Return
1,2,15,348,7760	85.31%
1,1,2,22,275,2500	85.33%
1,2,21,335,7000	87.68%
1,2,20,390,7000	90.85%
1,2,21,400,7000	92.44%
1,1,4,33,175,325	92.64%
1,2,22,422,7000	94.92%

Pick 8

Paytable	Return
1,12,112,1500,8000	84.17%
1,2,10,40,500,5000*	84.57%
2,12,98,1450,8000	88.20%
2,12,98,1550,10000	90.67%
2,12,98,1652,10000	92.31%
1,30,70,250,350*	92.62%
2,13,100,1670,10000	94.90%

Pick 9

Paytable	Return
1,5,40,400,1000,5000*	84.25%
1,3,47,352,4700,9000	84.87%
1,6,40,300,4700,9000	87.57%
1,6,44,300,4700,10000	89.93%
1,6,44,335,4700,10000	92.00%
1,1,8,65,178,325,375	92.66%
1,6,44,362,4700,10000	93.60%

*Catching zero pays 1.

Pick 10

Paytable	Return
5,25,120,500,2000,10000*	86.43%
3,28,140,1000,4800,10000	86.72%
5,21,142,1000,4000,10000	88.80%
5,23,132,1000,4500,10000	89.79%
5,24,142,1000,4500,10000	92.55%
1,1,2,36,115,225,300,400	92.69%
5,24,146,1000,4500,10000	93.20%

Caveman Keno 8X

In Caveman Keno, the game picks three numbers at random that the player didn't pick and marks them with eggs. Then the game draws 20 numbers from all 80. If the 20-number ball draw matches two of the egg numbers, as shown by the egg hatching, the player wins a 4X multiplier. If all three egg numbers match the ball draw, the player wins an 8X multiplier.

Pick 2

Paytable	Return
1,3	82.39%
10	86.41%
11	95.05%

Pick 3

Paytable	Return
2,23	84.86%
3,14	87.48%
3,15	89.42%
3,16	91.36%
3,17	93.30%
3,18	95.25%

Pick 4

Paytable	Return
1,5,60	86.59%
1,5,65	88.68%
1,5,69	90.35%
1,5,74	92.43%
1,5,78	94.10%
1,5,81	95.36%

Pick 5

Paytable	Return
1,2,4,200	87.42%
1,2,11,75	88.34%
1,2,12,80	90.43%
1,2,13,80	92.09%
1,2,14,85	94.18%
1,2,15,75	94.98%

Pick 6

Paytable	Return
2,5,46,750	88.43%
2,6,60,150	88.17%
2,6,65,150	90.24%
2,6,70,150	92.31%
2,6,74,150	93.96%
2,6,77,150	95.20%

Pick 7

Paytable	Return
1,3,14,230,1000	88.30%
1,3,12,225,2000	88.57%
1,3,14,255,1000	90.68%
1,3,14,270,1000	92.11%
1,3,14,290,1000	94.01%
1,3,14,300,1000	94.96%

Pick 8

Paytable	Return
1,2,5,62,90,1000	88.11%
1,2,4,46,450,2000	88.51%
1,2,5,65,155,1000	90.36%
1,2,5,68,200,1000	92.20%
1,2,5,74,200,1000	94.06%
1,2,5,77,200,1000	94.98%

Pick 9

Paytable	Return
1,5,35,225,1500,2000	87.88%
1,6,50,90,250,1000	88.36%
1,6,50,105,500,1000	90.49%
1,6,50,125,500,1000	92.00%
1,6,50,152,500,1000	94.03%
1,6,50,166,500,1000	95.08%

Pick 10

Paytable	Return
1,4,9,80,400,1500,2000	87.33%
1,5,10,60,230,500,1000	88.20%
1,5,10,70,250,500,1000	90.60%
1,5,10,77,250,500,1000	92.04%
1,5,10,87,250,500,1000	94.10%
1,5,10,90,275,500,1000	95.14%

Caveman Keno 10X

Caveman Keno 10X is the same thing as the 8X version, except a win is multiplied by 10 if all three eggs hatch. Following are all paytables I'm aware of.

Pick 2

Paytable	Return
1,3	83.75%
1,4	92.52%

Pick 3

Paytable	Return
2,23	86.07%
2,24	88.04%
2,26	91.98%

Pick 4

Paytable	Return
1,5,60	87.79%
1,5,67	90.75%
1,5,72	92.86%
1,5,75	94.13%

Pick 5

Paytable	Return
1,2,4,200	88.67%
1,2,5,215	91.64%
1,2,6,215	93.32%
1,2,6,230	94.62%

Pick 6

Paytable	Return
2,5,46,750	89.53%
2,5,49,800	91.63%
2,5,54,780	93.38%
2,5,56,800	94.55%

Pick 7

Paytable	Return
1,3,12,225,2000	89.62%
1,3,12,245,2000	91.54%
1,3,13,255,2000	93.68%
1,3,14,255,2000	94.85%

Pick 8

Paytable	Return
1,2,4,46,450,2000	89.63%
1,2,4,48,500,2000	91.28%
1,2,4,52,550,2000	93.55%
1,2,4,56,550,2000	94.80%

Pick 9

Paytable	Return
1,5,35,225,1500,2000	88.79%
1,5,35,255,1500,2000	91.07%
1,5,40,235,1500,2000	93.35%
1,5,40,250,1500,2000	94.49%

Pick 10

Paytable	Return
1,4,9,80,400,1500,2000	88.29%
1,4,9,100,300,1500,2000	90.76%
1,4,10,100,325,1500,2000	92.72%
1,4,11,100,325,1500,2000	94.26%

Caveman Keno Plus

Caveman Keno Plus is the same as Caveman Keno 8X, except a win with at least two eggs hatched draws three extra numbers. Following are the known paytables.

Pick 2

Paytable	Return
1,3	86.30%

Pick 3

Paytable	Return
2,20	87.83%
2,21	90.22%
2,22	92.60%

Pick 4

Paytable	Return
1,4,50	87.89%
1,4,54	90.15%
1,4,55	90.71%
1,4,57	91.84%
1,4,61	94.10%

Pick 5

Paytable	Return
1,2,5,88	88.15%
1,2,5,105	90.29%
1,2,5,110	90.91%
1,2,6,100	91.85%
1,2,6,118	94.11%

Pick 6

Paytable	Return
1,5,55,500	88.16%
1,5,58,500	89.93%
1,5,60,500	91.11%
1,6,53,500	91.89%
1,6,57,500	94.25%

Pick 7

Paytable	Return
1,3,10,110,1000	88.20%
1,3,11,112,1000	90.13%
1,3,12,108,1000	91.16%
1,3,13,104,1000	92.19%
1,3,14,106,1000	94.11%

Pick 8

Paytable	Return
1,2,4,20,200,2000	87.98%
1,2,4,21,250,2000	90.18%
1,2,4,22,108,1000	84.73%
1,2,4,25,250,2000	92.07%
1,2,5,22,250,2000	94.05%

Pick 9

Paytable	Return
2,4,15,120,500,2000	88.13%
2,4,16,125,500,2000	89.88%
2,4,17,125,500,2000	90.99%
2,4,18,125,500,2000	92.11%
2,4,19,132,500,2000	94.10%

Pick 10

Paytable	Return
1,2,10,60,250,1000,2000	87.93%
1,2,11,60,250,1000,2000	90.13%
1,2,11,63,250,1000,2000	91.14%
1,2,12,59,250,1000,2000	92.00%
1,2,12,65,250,1000,2000	94.02%

Cleopatra Keno

Cleopatra Keno plays like standard spot keno, except if the 20th ball drawn contributes to a win, the player wins 12 free games with a 2X multiplier. Following are the known paytables.

Pick 3

Paytable	Return
3,16	88.27%
3,17	90.19%
3,18	92.11%
3,19	94.03%

Pick 4

Paytable	Return
1,5,31	88.04%
1,5,35	90.10%
1,5,39	92.16%
1,5,43	94.22%
1,5,45	95.25%

Pick 5

Paytable	Return
3,21,220	88.34%
3,22,220	89.99%
3,23,225	92.08%
3,24,230	94.17%
3,24,240	95.05%

Pick 6

Paytable	Return
2,5,30,375	88.27%
2,5,34,375	90.28%
2,5,37,390	92.10%
2,5,40,410	94.03%
2,5,42,410	95.03%

Pick 7

Paytable	Return
1,3,6,80,500	88.01%
1,3,6,95,500	90.14%
1,3,7,97,500	92.09%
1,3,7,111,500	94.08%
1,3,7,118,500	95.07%

Pick 8

Paytable	Return
3,10,50,180,1000	87.99%
3,10,56,180,1000	90.15%
3,10,60,200,1000	92.08%
3,10,65,210,1000	94.12%
3,10,65,250,1000	95.09%

Pick 9

Paytable	Return
2,5,10,65,200,1000	88.44%
2,5,10,80,200,1000	90.02%
2,5,10,100,200,1000	92.14%
2,5,12,100,200,1000	94.19%
2,5,12,108,200,1000	95.04%

Pick 10

Paytable	Return
1,3,5,25,100,1000,2000	88.58%
1,3,5,25,150,1000,2000	90.01%
1,3,5,30,160,1000,2000	92.00%
1,3,5,35,175,1000,2000	94.13%
1,3,5,35,206,1000,2000	95.02%

Extra Draw Keno

In Extra Draw Keno, the player usually has the option of buying three extra balls after a win for the cost of his initial wager. The option isn't offered if it's an extremely poor value. For example, if the expected return on the extra balls were only, say, 10%, it wouldn't be offered. I'm sure they would offer it at 80%. Where the casino draws the line between poor and worthwhile value, I don't know. Also, be warned that some keno games in Amsterdam offer only two extra balls.

The strategy column of the following tables shows the number of catches after 20 balls where the odds favor paying for the extra three balls. The return column shows the ratio of the expected return to the initial bet, assuming optimal player strategy.

Pick 3

Paytable	Strategy	Return
3,29	2	86.03%
3,30	2	88.11%
3,31	2	90.19%
3,32	2	92.27%
3,33	2	94.35%

Pick 4

Paytable	Strategy	Return
1,12,31	2	85.07%
1,12,36	2,3	87.65%
1,12,40	2,3	89.88%
1,12,44	2,3	92.12%
1,12,47	2,3	93.79%

Pick 5

Paytable	Strategy	Return
1,4,13,100	3,4	87.11%
1,4,14,100	3,4	89.07%
1,4,14,100	3,4	89.07%
1,4,16,100	3,4	92.99%
1,4,15,125	3,4	94.51%

Pick 6

Paytable	Strategy	Return
4,7,25,200	4,5	87.66%
4,7,29,200	4,5	89.94%
4,8,28,200	4,5	91.95%
4,7,36,200	4,5	93.93%
4,7,38,200	4,5	95.07%

Pick 7

Paytable	Strategy	Return
2,6,15,34,400	4,5,6	87.43%
2,7,13,28,400	3,5,6	88.23%
2,7,14,32,400	3,4,5,6	90.40%
2,7,15,36,400	3,4,5,6	92.81%
2,7,15,43,400	3,4,5,6	94.07%

Pick 8

Paytable	Strategy	Return
1,4,11,20,75,800	4,5,6,7	86.97%
1,5,10,15,55,800	3,4,6,7	87.28%
1,5,10,17,85,800	3,4,5,6,7	89.74%
1,5,10,20,100,800	3,4,5,6,7	92.13%
1,5,10,20,130,800	3,4,5,6,7	93.58%

Pick 9

Paytable	Strategy	Return
1,3,5,10,31,240,2000	5,6,7,8	87.64%
1,3,5,12,30,240,2000	5,6,7,8	89.62%
1,3,5,13,35,240,2000	5,6,7,8	91.53%
1,3,5,14,41,250,2000	5,6,7,8	93.71%
1,3,5,14,44,300,2000	5,6,7,8	94.81%

Pick 10

Paytable	Strategy	Return
3,5,8,14,40,500,2500	3,6,7,8,9	84.72%
3,5,9,15,50,500,2500	3,5,6,7,8,9	86.69%
3,5,10,16,55,500,2500	3,5,6,7,8,9	89.43%
3,5,10,20,75,500,2500	3,5,6,7,8,9	92.09%
3,5,10,20,100,500,2500	3,5,6,7,8,9	93.28%

Power Keno

Power Keno plays like standard spot keno, except if the 20th ball drawn contributes to a win, that win is quadrupled.

Pick 2

Paytable	Return
11	85.98%
12	93.80%

Pick 3

Paytable	Return
2,26	88.39%
2,27	90.40%
2,28	92.41%
3,20	94.35%

Pick 4

Paytable	Return
1,5,60	88.41%
1,5,61	88.90%
1,6,50	89.78%
1,7,40	91.14%
1,6,55	92.23%
1,6,59	94.19%

Pick 5

Paytable	Return
1,2,11,75	89.25%
2,20,230	88.99%
1,2,10,100	90.13%
1,2,12,75	91.18%
1,2,11,100	92.07%
1,2,12,100	94.00%

Pick 6

Paytable	Return
2,7,30,125	88.93%
1,5,65,500	89.12%
2,7,31,150	90.08%
2,8,24,160	91.10%
2,7,35,160	92.49%
2,8,30,150	94.10%

Pick 7

Paytable	Return
1,4,12,75,200	88.35%
1,2,14,140,1250	88.97%
1,4,14,60,200	89.29%
1,4,12,90,200	90.44%
1,4,14,75,200	91.37%
1,5,10,75,200	93.68%

Pick 8

Paytable	Return
1,2,6,17,120,500	88.51%
2,7,41,600,2500	89.06%
1,2,6,21,100,500	89.65%
1,2,6,24,100,500	91.00%
1,2,6,26,100,500	91.90%
1,2,7,25,80,500	94.00%

Pick 9

Paytable	Return
2,5,14,55,200,1000	88.53%
1,4,20,150,1000,3750	88.81%
2,4,16,90,200,1000	89.24%
2,5,15,60,200,1000	90.22%
2,5,16,60,200,1000	91.31%
2,6,14,50,200,1000	93.63%

Pick 10

Paytable	Return
1,4,8,24,100,500,1000	88.67%
1,2,7,45,500,1750,5000	89.25%
1,4,8,25,125,500,1000	89.75%
1,4,8,25,150,500,1000	90.49%
1,4,9,25,120,500,1000	91.78%
1,4,10,25,120,500,1000	93.96%

Triple Power

Triple Power Keno plays like Power Keno, except if the 20th ball contributes to a win, that win is tripled, plus the player gets three extra balls.

Pick 2

Paytable	Return
1,6	86.49%

Pick 3

Paytable	Return
2,27	88.75%
2,28	90.77%
3,21	93.68%

Pick 4

Paytable	Return
1,6,48	88.62%
1,6,51	90.24%
1,7,40	90.50%
1,6,55	92.38%
1,6,58	94.00%

Pick 5

Paytable	Return
1,2,10,85	88.34%
1,2,10,100	90.44%
1,2,12,75	91.07%
1,2,11,100	92.50%
1,2,12,100	94.56%

Pick 6

Paytable	Return
2,6,31,125	88.50%
2,6,33,150	90.68%
2,8,19,150	91.08%
2,7,28,150	92.18%
2,8,24,150	94.32%

Pick 7

Paytable	Return
1,4,11,54,200	88.62%
1,4,12,55,200	90.56%
1,4,12,59,200	91.33%
1,4,12,65,200	92.49%
1,5,8,65,250	94.35%

Pick 8

Paytable	Return
1,2,6,12,90,500	88.55%
1,2,5,20,100,500	90.31%
1,2,6,16,95,500	91.22%
1,2,5,24,95,500	92.44%
1,2,5,27,92,500	94.08%

Pick 9

Paytable	Return
2,5,10,32,150,1000	88.62%
2,3,13,80,200,1000	90.41%
2,5,10,42,200,1000	91.27%
2,5,10,50,175,1000	92.45%
2,5,12,45,180,1000	94.35%

Pick 10

Paytable	Return
1,4,6,16,50,400,1000	88.40%
1,3,7,25,120,500,1000	90.59%
1,4,7,14,72,350,1000	91.18%
1,4,7,15,90,250,1000	92.35%
1,3,9,27,80,400,1000	94.32%

19

Video Poker

Video poker is one of the two traditional casino games (along with blackjack) to which most advantage players used to gravitate. Even with the best paytables and perfect play, the casino almost always holds a mathematical advantage, before considering the value of player reward points. Speed demons can rip through *thousands* of hands per hour, which helps them get to the long term much faster than at blackjack. And comps can add up just as quickly.

Unfortunately, the heyday of video poker paydays is a thing of the past. Paytables have been reined in, comps have been tightened, and though backoffs aren't as common at VP as blackjack, they do happen. Over the past several years, the ranks of professional video poker players have shrunk dramatically. That's not to say opportunities don't still exist. They do. For low stakes, bar promotions, sign-up bonuses, and potentially positive progressives are common. And even for higher stakes, opportunities arise with promotions and progressives, as well as some casinos' willingness to offer beatable games when factoring in players club value, comps, and sometimes, outright mistakes.

In any event, video poker remains a very popular game, primarily because it's so much fun. In fact, it can be addicting. But if you follow my advice in this chapter, you'll make the most of your video poker play.

Video poker advice can be broken down into four steps:

1. Bet enough to maximize the win per coin bet for a royal flush.
2. Shop around for a good game and paytable.
3. "… knowin' what to throw away and knowin' what to keep." — Kenny Rogers.
4. Milk player-reward programs for as much as possible.

I know, easier said than done. This chapter covers these four steps in as much detail as a short book can, with particular emphasis on steps 2 and 3.

Bet Max Coins

On most video poker machines, the win for a royal depends on the number of coins bet, as follows.

Coins Bet	Royal Flush Pays
1	250
2	500
3	750
4	1000
5	4000

If you divide the win by coins bet, you get 250 for playing 1-4 coins bet and 800 for 5. For that reason, you should always make a max-coin bet on such machines. Not doing so will be a mistake that costs you about 1.36% in expected value. Be warned that some games don't have this max-bet incentive and on others, it doesn't kick in until 10 coins are bet.

As long as I'm on this topic, there are a slew of new video

poker games where you make an extra non-refundable bet to enable some kind of bonus feature. (Non-refundable means that the fee isn't part of the original bet. For example, in Ace on the Deal, the player can pay an extra coin, on top of a five-coin bet, to guarantee he gets at least one ace on the deal. It's completely non-refundable.) The maker of these games, IGT/ Action Gaming, always seems to reward the player for making this extra bet with a higher return, ranging from about 0.01% to 0.5%. Thus, my advice about making a max-coin bet applies to these games too.

In addition, on games that allow multiple hands, my advice applies to the bet on the hand itself, not the number of hands played. How many hands you play, when given the choice, doesn't matter for purposes of the expected return.

Shopping for Paytables

You can't shop around for a good paytable if you don't know what a good paytable is. Most recreational players know what games they like, so they plop down and play without even checking the paytable. However, the house edge depends very much on how much each hand returns. For example, playing 6/5 Jacks or Better at the Las Vegas airport has a house edge 11 times higher than the liberal 9/6 paytable.

On most hands, the pays will be the same from one machine to the next. On games with no wild cards, they tend to differ from one another on only two hands, the flush and full house. As a rule of thumb, for every coin the win on a flush or full house is lowered (based on one coin bet), the return goes down by 1.1%.

That said, the following tables present the expected

return, assuming optimal strategy, for all the major video poker games and available paytables. Wins are per coin bet, based on a maximum-coin bet. Odds are generally expressed as an "expected return" on machines and "expected value" with table games. I'm sticking with that convention. If you want the house edge, just subtract the expected return from 100%.

The way I suggest you use this information is to compare the return of any game and paytable you consider playing to the tables in this chapter. If you play a lot, you'll quickly find that you see the same paytables over and over and will know the good from the bad on sight. Until then, use these tables. Your long-term results in video poker are highly dependent on the paytables you play.

Jacks or Better

Royal Flush	800	800	800	800	800	800
Straight Flush	50	50	50	50	50	50
4-of-a-Kind	25	25	25	25	25	25
Full House	9	9	8	8	7	6
Flush	6	5	6	5	5	5
Straight	4	4	4	4	4	4
3-of-a-Kind	3	3	3	3	3	3
Two Pair	2	2	2	2	2	2
Pair	1	1	1	1	1	1
Total	99.54%	98.45%	98.39%	97.30%	96.15%	95.00%

Bonus Poker

Royal Flush	800	800	800	800
Straight Flush	50	50	50	50
4 Aces	80	80	80	80
4 2s-4s	40	40	40	40
4 5s-Ks	25	25	25	25
Full House	8	7	6	10
Flush	5	5	5	8
Straight	4	4	4	5
3-of-a-Kind	3	3	3	3
Two Pair	2	2	2	1
Jacks or Better	1	1	1	1
Total	**99.17%**	**98.01%**	**96.87%**	**94.18%**

Bonus Poker Deluxe

Royal Flush	800	800	800	800	800
Straight Flush	50	50	50	50	50
4-of-a-Kind	80	80	80	80	80
Full House	9	8	8	7	6
Flush	6	6	5	5	5
Straight	4	4	4	4	4
3-of-a-Kind	3	3	3	3	3
Two Pair	1	1	1	1	1
Pair	1	1	1	1	1
Total	**99.64%**	**98.49%**	**97.40%**	**96.25%**	**95.36%**

Double Bonus

Royal Flush	800	800	800	800	800	800	800	800	800
Straight Flush	50	50	50	50	50	50	50	50	50
4 Aces	160	160	160	160	160	160	160	160	160
4 2s-4s	80	80	80	80	80	80	80	80	80
4 5s-Ks	50	50	50	50	50	50	50	50	50
Full House	10	9	10	9	9	9	8	9	7
Flush	7	7	7	6	7	6	6	5	5
Straight	5	5	4	5	4	4	5	4	4
3-of-a-Kind	3	3	3	3	3	3	3	3	3
Two Pair	1	1	1	1	1	1	1	1	1
Jacks or Better	1	1	1	1	1	1	1	1	1
Total	100.17%	99.11%	98.81%	97.81%	97.74%	96.38%	96.73%	95.27%	93.11%

Double Double Bonus

Royal Flush	800	800	800	800	800	800	800
Straight Flush	50	50	50	50	50	50	50
4 Aces + 2-4	400	400	400	400	400	400	400
4 2s-4s + A-4	160	160	160	160	160	160	160
4 Aces + 5-K	160	160	160	160	160	160	160
4 2s-4s + 5-K	80	80	80	80	80	80	80
4 5s-Ks	50	50	50	50	50	50	50
Full House	10	9	8	9	8	7	6
Flush	6	6	6	5	5	5	5
Straight	4	4	4	4	4	4	4
3-of-a-Kind	3	3	3	3	3	3	3
Two Pair	1	1	1	1	1	1	1
Jacks or Better	1	1	1	1	1	1	1
Total	100.07%	98.98%	97.89%	97.87%	96.79%	95.71%	94.66%

Triple Double Bonus

Royal Flush	800	800	800	800	800	800
Straight Flush	50	50	50	50	50	50
4 Aces + 2-4	800	800	800	800	800	800
4 2s-4s + A-4	400	400	400	400	400	400
4 Aces + 5-K	160	160	160	160	160	160
4 2s-4s + 5-K	80	80	80	80	80	80
4 5s-Ks	50	50	50	50	50	50
Full House	9	9	9	8	7	6
Flush	7	6	5	5	5	5
Straight	4	4	4	4	4	4
3-of-a-Kind	2	2	2	2	2	2
Two Pair	1	1	1	1	1	1
Jacks or Better	1	1	1	1	1	1
Total	99.58%	98.15%	97.02%	95.97%	94.92%	93.87%

Triple Bonus Poker Plus

Royal Flush	800	800	800	800
Straight Flush	100	100	100	100
4 Aces	240	240	240	240
4 2-4	120	120	120	120
4 5-K	50	50	50	50
Full House	9	8	7	6
Flush	5	5	5	5
Straight	4	4	4	4
3-of-a-Kind	3	3	3	3
Two Pair	1	1	1	1
Jacks or Better	1	1	1	1
Total	99.80%	98.73%	97.67%	96.62%

Super Double Bonus

Royal Flush	800	800	800	800
Straight Flush	80	80	80	80
4 Aces	160	160	160	160
4 Js-Ks	120	120	120	120
4 2s-4s	80	80	80	80
4 5s-10s	50	50	50	50
Full House	9	8	7	6
Flush	5	5	5	5
Straight	4	4	4	4
3-of-a-Kind	3	3	3	3
Two Pair	1	1	1	1
Jacks or Better	1	1	1	1
Total	99.69%	98.69%	97.77%	96.87%

Ace$ Bonus Poker

Royal Flush	800	800	800	800	800	800	800
Straight Flush	50	50	50	50	50	50	50
ACE$- or -ACE$	800	800	800	800	800	800	800
Four A	80	80	80	80	80	80	80
Four 2-4	40	40	40	40	40	40	40
Four 5-K	25	25	25	25	25	25	25
Full House	8	7	6	10	10	9	10
Flush	5	5	5	8	6	6	8
Straight	4	4	4	6	4	4	4
3-of-a-Kind	3	3	3	3	2	2	3
Two Pair	2	2	2	1	2	2	1
Pair	1	1	1	1	1	1	1
Total	99.41%	98.26%	97.11%	96.01%	95.36%	94.21%	93.05%

Super Double Double Bonus Poker

Royal Flush	800	800	800	800	800
Straight Flush	50	100	50	100	50
Four Aces + 2-4	400	400	400	400	400
Four Aces + J-K	320	320	320	320	320
Four 2s-4s + A-4	160	160	160	160	160
Four Js-Ks + J-A	160	160	160	160	160
Four Aces	160	160	160	160	160
Four 2s-4s + 5-K	80	80	80	80	80
Four 5s-Ks	50	50	50	50	50
Full House	8	7	7	6	6
Flush	5	5	5	5	5
Straight	4	4	4	4	4
3-of-a-Kind	3	3	3	3	3
Two Pair	1	1	1	1	1
Jacks or Better	1	1	1	1	1
Total	99.69%	99.17%	98.61%	98.25%	97.69%

Shockwave

Royal Flush	800	800	800	800	800	800
Straight Flush	100	100	100	100	100	100
4-of-a-Kind - Shockwave mode	800	800	800	800	800	800
4-of-a-Kind - Regular mode	25	25	25	25	25	25
Full House	12	11	10	10	9	9
Flush	8	8	8	7	6	6
Straight	5	5	5	5	5	4
3-of-a-Kind	3	3	3	3	3	3
Two Pair	1	1	1	1	1	1
Pair	1	1	1	1	1	1
Total	99.55%	98.44%	97.34%	95.72%	93.23%	91.77%

White Hot Aces

Royal Flush	800	800	800	800
Straight Flush	80	80	80	80
4 Aces	240	240	240	240
4 2-4	120	120	120	120
4 5-K	50	50	50	50
Full House	9	8	7	6
Flush	5	5	5	5
Straight	4	4	4	4
3-of-a-Kind	3	3	3	3
Two Pair	1	1	1	1
Jacks or Better	1	1	1	1
Total	**99.57%**	**98.50%**	**97.44%**	**96.39%**

Deuces Wild

Royal Flush	800	800	800	800	800	800	800	800	800	800	800
Four Deuces	200	200	200	200	200	200	200	200	200	200	200
Wild Royal	25	25	25	25	20	25	20	20	25	20	25
5-of-a-Kind	15	15	16	15	12	15	12	12	16	10	15
Straight Flush	9	11	10	10	9	9	10	9	13	8	10
4-of-a-Kind	5	4	4	4	5	4	4	4	4	4	4
Full House	3	4	4	4	3	4	4	4	3	4	3
Flush	2	3	3	3	2	3	3	3	2	3	2
Straight	2	2	2	2	2	2	2	2	2	2	2
3-of-a-Kind	1	1	1	1	1	1	1	1	1	1	1
Total	**100.76%**	**99.96%**	**99.73%**	**99.42%**	**98.94%**	**98.91%**	**97.58%**	**97.06%**	**96.77%**	**95.96%**	**94.82%**

Bonus Deuces

Royal Flush	800	800	800	800	800	800	800
Four Deuces w/Ace	400	400	400	400	400	400	400
Four Deuces	200	200	200	200	200	200	200
Wild Royal	25	25	25	25	25	25	25
Five Aces	80	80	80	80	80	80	80
Five 3-5	40	40	40	40	40	40	40
Five 6-K	20	20	20	20	20	20	20
Straight Flush	10	9	8	13	10	12	10
4-of-a-Kind	4	4	4	4	4	4	4
Full House	4	4	4	3	3	3	3
Flush	3	3	3	3	3	2	2
Straight	1	1	1	1	1	1	1
3-of-a-Kind	1	1	1	1	1	1	1
Total	99.86%	99.45%	99.06%	98.80%	97.36%	96.22%	95.34%

Joker Poker (Kings or Better)

Royal Flush	800	800	940	800	940	800	800
5-of-a-Kind	200	200	200	200	200	200	200
Wild Royal	100	100	100	100	100	100	100
Straight Flush	50	50	50	50	50	50	40
4-of-a-Kind	20	18	17	17	15	15	20
Full House	7	7	7	7	7	7	5
Flush	5	5	5	5	5	5	4
Straight	3	3	3	3	3	3	3
3-of-a-Kind	2	2	2	2	2	2	2
Two Pair	1	1	1	1	1	1	1
Kings or Better	1	1	1	1	1	1	1
Total	100.65%	98.94%	98.44%	98.09%	96.74%	96.38%	95.46%

Joker Poker (Two Pair or Better)

Royal Flush	1000	800	100	100	100
5-of-a-Kind	100	800	800	1000	1000
Wild Royal	50	100	100	100	100
Straight Flush	50	100	100	40	40
4-of-a-Kind	20	16	16	15	15
Full House	10	8	8	8	7
Flush	6	5	5	5	5
Straight	5	4	4	4	4
3-of-a-Kind	2	2	2	2	2
Two Pair	1	1	1	1	1
Return	**99.92%**	**98.59%**	**97.19%**	**93.90%**	**92.66%**

Joker Poker (Aces or Better)

Natural Royal Flush	800	800	1000	800	1000	800
Five-of-a-Kind	200	200	200	200	200	200
Wild Royal Flush	100	100	100	100	100	100
Straight Flush	50	50	50	50	50	50
Four-of-a-Kind	20	20	20	20	20	20
Full House	9	8	7	7	6	6
Flush	5	5	5	5	5	5
Straight	3	3	3	3	3	3
Three-of-a-Kind	2	2	2	2	2	2
Two Pair	1	1	1	1	1	1
Aces or Better	1	1	1	1	1	1
Return	**98.42%**	**96.87%**	**95.82%**	**95.32%**	**94.27%**	**93.78%**

Playing Strategy

Optimal video poker strategy is different for every game and paytable. On most games, you can get within 0.05% of the optimal return with a simplified strategy. That's the case with the following strategy for 9/6 Jacks or Better. This strategy can be used with little cost in mistakes for any Jacks or Better, Bonus Poker, or Bonus Poker Deluxe game. Note that the chart lists the cards you should hold in order of preference.

9/6 Jacks or Better Strategy

Royal flush, straight flush, 4-of-a-kind
Four to a royal flush
Full house, flush, straight, 3-of-a-kind
Four to a straight flush
Two pair
High pair (jacks or higher)
Three to a royal flush
Four to a flush
Low pair (10s or lower)
Four to an outside straight
Three to a straight flush (type 1, see following page)
AKQJ unsuited
Two suited high cards
Four to an inside straight with three high cards
Three to a straight flush (type 2, see following page)
KQJ unsuited
QJ unsuited
JT suited
KQ, KJ unsuited
QT suited
AK, AQ, AJ unsuited
KT suited
One high card
Three to a straight flush (Type 3, see following page)
Discard everything

Types of Three to a Straight Flush

Three to a straight flush is a common and tricky play in video poker. The value depends on the number of high cards, which are good, and the number of gaps, which are bad. Conveniently, the value of a high card is nearly offset by the cost of a gap. That said, here are the three types of straight flush holds.

Type 1: The number of high cards is greater or equal to the number of gaps. For example, a suited Q,10,9, with two junk cards (one high card and one gap).

Type 2: The number of gaps less high cards equals one, 2-3-4, and all ace low. For example, a suited Q,10,8 and two junk cards (one high card and two gaps).

Type 3: Two gaps and no high cards. For example, a suited 7,5,3 and two junk cards.

To use this or any similar type of video poker strategy, look up all viable ways to play the hand and go with the one listed highest. If a hand isn't listed, like suited A,10, it should never be played (holding the ace alone is better). For example, consider the following hand.

10♥ 3♣ Q♥ 10♠ A♥

Common sense should tell us the two best plays are the three to a royal and the pair of tens. Sure enough, the strategy shows three to a royal in the number seven spot and the low pair in the number nine spot. Since the three to a royal is listed higher, that's the play.

NSU Deuces Wild Strategy

The next strategy is optimized for 25-16-10-4-4-3 Deuces Wild, known as "Not So Ugly Ducks" or just "NSU," among serious players, and has an optimal return of 99.73%. The Deuces Wild strategy in the first edition of this book was based on "Full Pay Deuces Wild," with a return of 100.76%, but that game is simply too difficult to find any longer to dignify it with a published strategy in this book.

For hands with four deuces, hold them all. It's optional to hold the singleton.

Hands with 3 deuces
> Wild royal
> 5-of-a-kind
> Three deuces

Hands with 2 deuces
> Pat 4-of-a-kind to wild royal
> Four to a royal flush
> Four to a straight flush: 0 or 1 gap except 34, 35, 46
> Two deuces

Hands with 1 deuce
> Straight flush to wild royal
> Four to a royal flush
> Flush to 4K
> Four to a straight flush: 0 or 1 gap
> Straight
> Four to a straight flush: 2 gaps
> 3-of-a-kind
> Four to a straight flush: A low

Three to a royal flush: J-K high
Three to a straight flush: 0 gaps (67 to 9T)
Three to a royal flush: A high
Three to a straight flush: 1 gap (57 to 9J), 45, 56
Four to a straight: 0 gaps (567 to TJQ)
One deuce

Hands with 0 deuces
Royal flush
Four to a royal flush
3-of-a-kind through straight flush
Four to a straight flush
Three to a royal flush
Four to a flush
Two pair
Three to a straight flush: 0 gaps (567 to 9TJ)
Pair
Four to a straight: 0 gaps (4567 to TJQK)
Three to a straight flush: 1 or 2 gaps, 345, 456
Two to a royal flush: J or Q high
Three to a straight flush: A low
Four to a straight: 1 gap, 3456
Two to a royal flush: K high
Toss everything

Player-Reward Programs

For any form of gambling, you should play with a players card, but this is especially true for video poker. Today, professional video poker players rarely play games with a theoretical return over 100%, because what few that still exist are at low denominations. Instead, sharp players are willing to take a theoretical loss on their straight-up play, because they get back more in the form of free-play and other comps. Not to use a players card, no matter your skill level, is throwing money away.

The way the programs usually work is you get a certain number of points per dollar bet and it takes another number of points to redeem a dollar's worth of comps and, in some cases, free-play. If this is the case, the value of points is the number of points earned per dollar played divided by the points required to redeem a dollar.

For example, consider Station casinos on a 6X points day. On these special days, the video poker player gets 6 points per dollar bet. You can redeem 1,000 points for $1 worth of comps (I suspect most people redeem them in the restaurants). That makes the value of the points worth 6/1000 = 0.6%, assuming you value comps as much as cash.

Sometimes, the way the rules are expressed is an amount of play required to get a point. If this is the case, the points awarded per dollar is the inverse of the play required to get a point. For example, Caesars Entertainment's Total Rewards program requires $10 of video poker play to get one reward credit. That means points earned per dollar bet equals 1/10 = 0.1%. The best conversion rate seems to be 150 reward credits for a dollar in restaurant credit. In this case, the value of reward credits is 0.1/150 = 0.067%.

In the case of Caesars and many other properties, there is much more value in points than the value they can be cashed in for. As you can see, the cashable value of Total Rewards points is pretty low. However, almost every machine player will be bombarded with all kinds of offers and free play that don't require cashing in points. At many properties, it doesn't take much play to get offers of free rooms and sometimes more. With many programs, Total Rewards being a good example, you get much more by waiting for an invitation than begging for a discretionary comp (a comp written by a host).

How generous and stingy each property and program are vary all the time. Other players can be a great source of information on how much play is required to get what level of offers. Even if you don't understand the complicated rules of a reward program, sign up and use a card anyway, then wait and see what's offered.

Finally, I have no idea when you'll be reading this book, so the rules of the programs I use in this chapter as examples may change.

20

Gambling FAQ

What is the best game to play?

It depends on the rules of the game and how well you play it. Limiting the answer to popular games, assuming you play the optimal strategy and stick to all the best bets when given a choice, I'd narrow down the best games to the four in the following list. (The percentage shown is the element of risk of those games.)

- blackjack (6 decks, dealer stands on soft 17, double after split allowed, surrender allowed, re-splitting aces allowed): 0.25%
- craps (3-4-5X odds, laying the maximum odds allowed): 0.27%
- video poker (9/6 Jacks or Better): 0.46%
- Ultimate Texas Hold 'em: 0.53%

What is your favorite game?

My answer would be the same as the first question on the best game to play. However, the answer to the question about which game I find the most fun to play is pai gow (tiles). I dislike volatility and tiles offers a slow game with lots of pushes. It's also a challenging game to understand and play well. I find that other players are generally smart people and pleasant to play with.

Do bad players, in particular in blackjack, cause everybody else to lose?

No. While everyone remembers the time a bad player took the dealer's bust card and caused the whole table to lose, people tend to forget the times that a bad player saved the table. This practice of selective memory to support pre-existing beliefs is called "confirmation bias." In the long run, bad players are just as likely to help you as hurt you, so leave them alone.

What do you think of my betting system?

All betting systems are equally worthless. Not only can't a betting system overcome the house edge, it can't even dent it. If a betting system makes gambling more fun, be my guest. Just don't delude yourself that it will help in the long run.

How much do the casinos pay you to publish your awful and incorrect strategies?

The same amount they pay you to cast doubt on them.

What is your opinion on dice control?

For the benefit of those who don't understand the question, books, videos, and lessons allege that it's possible to beat the odds in craps with a careful toss that favors certain outcomes, namely lowering the probability of throwing a 7 to less than 1 in 6. I'm firmly in the skeptics camp on this one. I have yet to see any credible evidence leading me to believe that anyone can consistently influence the dice. There's much more money to be made selling books and lessons on how to do it than actually doing it.

Which is your favorite casino in Las Vegas?

The casino that I feel offers the best odds and overall value is South Point.

Where do casinos put the loosest slots?

As a rule of thumb, the location makes no difference.

Why do you say not to take "even money" on a blackjack when the dealer has an ace up? It's a sure win!

There is a 69.1% chance the dealer doesn't have a black-jack and you'll win the full 3-2. $(1.5 * .691 = 103.7\%.)$ That's more than the 100% you get by taking even money. You've already established the fact that you're a gambler by playing in the first place. Don't suddenly become risk-averse and give up that 3.7% because you don't want to take a chance.

Casino [insert name here] is cheating. Can you please warn your readers about them? I know because [insert adjective-laden story about losing here].

This kind of accusation rarely comes with any evidence behind it. What rare times I get some actual numbers, the loss could easily be explained as ordinary bad luck. Never-theless, I have exposed cases of cheating at Internet casinos several times, starting with such accusations. So if you sus-pect a casino is cheating, please follow the scientific method before writing to me. First, make a hypothesis about how the casino is cheating; second, gather evidence to confirm or deny the hypothesis; and third, analyze the evidence. I'm happy to help with step 3.

If a ball landed in red the last 20 spins in roulette, what is the probability it will land in black the next spin?

The same as red, 47.37% on a double-zero wheel, 18 black numbers divided by 38 total numbers.

I think you're wrong. The odds of 21 reds in a row is $(18/38)^{21}$ = 1 in 6,527,290. The odds must overwhelmingly favor black.

Your calculation is correct, but it doesn't matter. That's the same probability of 20 reds followed by a black. The fact is the past doesn't matter in games of independent trials like roulette.

Why are you such a Debbie Downer when it comes to gambling? You take all the fun out of it.

I can only lead a horse to water. You don't have to drink it if you don't want to.

Index of Tables

Pai Gow

Pai Gow Poker

Slots

Sports Betting

Texas Hold 'Em Bonus

Three Card Poker

Ultimate Texas Hold 'Em

About the Author

Michael Shackleford was born and raised in southern California. In 1988, he graduated with a degree in math and economics from the University of California Santa Barbara. After that, he worked as a Social Security actuary for 10 years, calculating how changes in the law would cost or save the taxpayers over a 10-year period. He also published the first list of baby-name popularity, based on a nationwide sample, which is a popular feature of the Social Security website to this day.

In 1997, as a hobby, Mike created the Wizard of Odds, a non-profit website on how to get the best odds possible in the major casino games. In 2000, Mike left a good government job and changed his website to a for-profit model. The next year, he moved from Baltimore to Las Vegas to focus his attention on writing about gambling.

Michael lost much of his money in 2013 when the Laiki Bank in Cyprus failed and Web advertising revenue fell about 90%. That was a bad year. The next year, Mike sold his website and now focuses his energy on writing content for the new owner.

Other than helping gamblers around the world improve their odds in the casino, Mike likes to keep busy with numerous hobbies, including mountain climbing, bicycling, unicycling, inline skating, solving puzzles, and trivia challenges. He continues to live in Las Vegas where he's the husband of one and the father of three.

LasVegasAdvisor.com/Shop
For Other Great Gambling Books and Products

The 21st Century Card Counter
by Colin Jones

A journey through the inner world of card counting, teamwork, and the clandestine pursuit of a blackjack livelihood, this is also the only book on the market that describes a professional-level technique for beating the casinos with a two- or three-player approach, as opposed to single-player or big-team play.

Then One Day ...
by Chris Andrews

Betting on sporting events is a deep and rich culture of big money, complex arithmetic, and a language all its own. Squares (the public), wiseguys (also known as the "sharp money"), and bookies converge in the legal sports books—and *Then One Day* describes the vastly colorful scene.

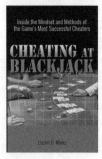

The Blackjack Insiders
by Andrew Uyal

A casino floor supervisor learns how to count cards from his shift manager. The two gambling executives perfect their card-counting technique, then use their inside knowledge of casino procedures to launch a no-holds-barred attack on 21 tables across the country. No book on blackjack has ever been written from such a deep-rooted behind-the-scenes perspective.

Cheating at Blackjack
by Dustin Marks

A legendary blackjack cheater who was never caught divulges decades-old secrets used by casino "crossroaders" to beat the game. *Cheating at Blackjack* gets into the minds and methods of the cheater, providing explanations of the techniques, illustrated by entertaining real-life accounts of how it was done.

To order visit LasVegasAdvisor.com/Shop or Call 800-244-2224

The Essentials of Casino Game Design
by Dan Lubin

The Essentials of Casino Game Design is the only manual in existence for designing, developing, marketing, and safeguarding a new casino table game, so that you can hit the jackpot by getting it installed on the floors of casinos.

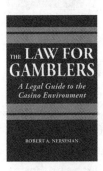

Law for Gamblers
by Robert A. Nersesian

Until recently, the casino industry, in its insatiable thirst for profits, has all but ignored the players' rights and interests. *The Law for Gamblers* puts the rights—and obligations—of players into historical, political, legal, and cultural perspective.

Gambling Wizards
by Richard W. Munchkin

Gambling Wizards takes you into the lives and minds of some of the most successful professional gamblers of all time, including famed sports bettors Billy Walters and Stan Tomchin, and horse-betting phenom Alan Wood.

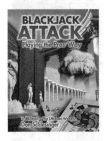

Blackjack Attack
by Don Schlesinger

This third and final edition of Don Schlesinger's essential blackjack tome, now in softcover, is the biggest and best ever—a massive 533 pages that contain more than 400 tables and charts. *Blackjack Attack* is one of the most important books available to blackjack aficionados, and has been praised by every prominent blackjack expert.

Keep up to Date on
All the Developments in Sports Betting

Go to GamblingWithAnEdge.com to access the blogs and podcasts of the world's top advantage players, including professional sports bettors who cover the subject from every conceivable angle. Shop lines on our live odds feed, get analyses of the latest online incentive offers, track the progress of newly legalized sports betting jurisdictions, and learn about new sports betting technologies, e.g., plans for the VBX betting exchange. Visit often for the latest information and opportunities in the explosion of sports betting nationwide.

Visit
LasVegasAdvisor.com
for all the latest on
gambling and Las Vegas

Free features include:

- Articles and ongoing updates on gambling.

- Tournament listings and articles

- Up-to-the-minute Las Vegas gambling-promotion announcements.

- Question of the Day offering in-depth answers to gambling- and Las Vegas-related queries.

- Active message boards with discussions on blackjack, sports betting, poker, and more!

Become a *Las Vegas Advisor* member and get our exclusive coupons and members-only discounts.

About Huntington Press

Huntington Press is a specialty publisher of Las Vegas- and gambling-related books and periodicals, including the award-winning consumer newsletter, Anthony Curtis' Las Vegas Advisor.

Huntington Press
3665 Procyon Street
Las Vegas, Nevada 89103